Grade 8

Discovery EDUCATION | SCIENCE TECHBOOK

California

Volume 2

Unit 3: Life's Unity and Diversity

Unit 4: Monitoring Biodiversity

To obtain permission(s) or for inquiries, submit a request to:
Discovery Education, Inc.
4350 Congress Street, Suite 700
Charlotte, NC 28209
800-323-9084
Education_Info@DiscoveryEd.com

ISBN 13: 978-1-68220-558-7

Printed in the United States of America.

1 2 3 4 5 LBC 28 27 26 25 24 B

Acknowledgments

Acknowledgment is given to photographers, artists, and agents for permission to feature their copyrighted material.

Cover and inside cover art: Ethan Daniels / Shutterstock.com

Letter to the Student

Dear Student,

See science in a whole new way! In this class, you'll be using California Science Techbook™. California Science Techbook is a science program developed by Discovery Education. The program is full of images, videos, Hands-On Activities, digital tools, reading passages, animations, and other activities. These resources will help you analyze and interpret data. You will solve problems and make connections between science and the world around you. California Science Techbook is made so you can work at your own pace and explore questions you may have about science. You'll even be able to see your progress in real time using the online Student Dashboard.

You will use this Student Edition to explore important ideas and record what you know and what you have learned. You'll also use it to make connections to the digital content in online Techbook. This will help you develop your own scientific understanding.

In each section of the Student Edition, you'll find QR codes. When you scan these codes, they'll take you to the online Science Techbook section you need. For instance, QR links throughout the book take you directly to important anchor or investigative phenomena videos and images. Once you are inside digital California Science Techbook, you can try some Explorations, Hands-On Activities, or Virtual Labs. All of them will help you explore the most important ideas in a concept. Enjoy this deep dive into the exciting world of science!

Sincerely,

The Discovery Education Science Team

Letter to the Parent/Guardian

Dear Parent/Guardian,

This year, your student will be using California Science Techbook™, a comprehensive science program developed by the educators and designers at Discovery Education.

California Science Techbook is an innovative program that offers engaging, real-world problems to help your student master key scientific concepts and act and think like a scientist. Students engage with interactive science instruction to analyze and interpret data, think critically, solve problems, and make connections across science disciplines. In addition, they experience dynamic content, explorations, videos, digital tools, Hands-On Activities and labs, and game-like activities that inspire and motivate scientific learning and curiosity.

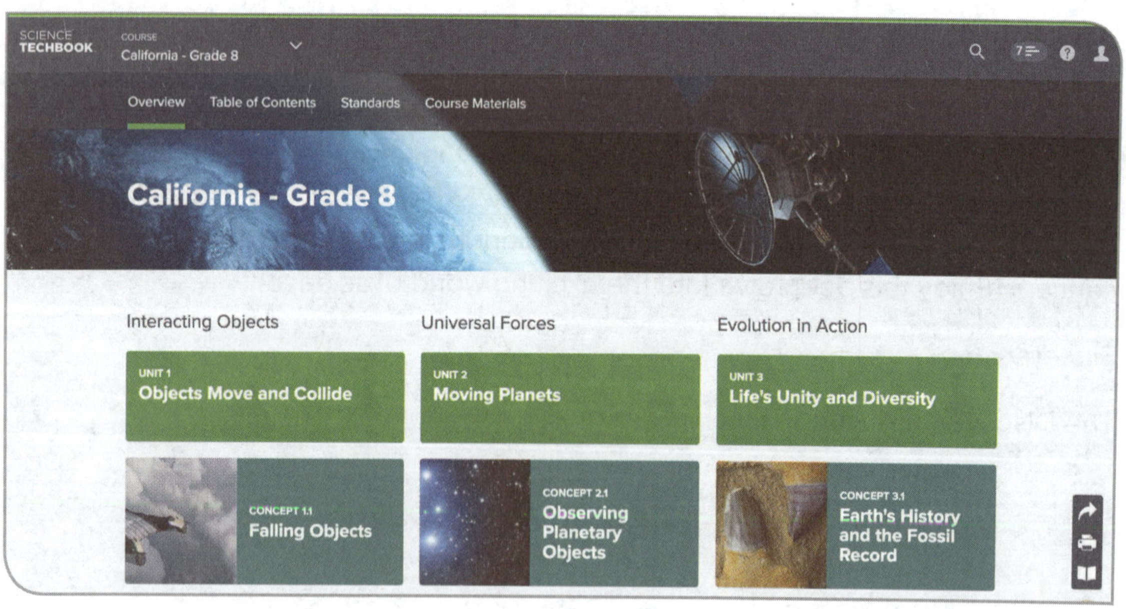

Within this Student Edition, you'll find QR codes that take you and your student to a corresponding section of Science Techbook online. Once in Techbook, students will have access to the Core Interactive Text of each concept, as well as thousands of resources and activities that build deep conceptual scientific understanding. Additionally, tools and features such as the Interactive Glossary and text-to-speech functionality allow Science Techbook to target learning for students of a variety of abilities.

To use the QR codes, you'll need to download a free QR reader. Readers are available for phones, tablets, laptops, desktops, and virtually any device in between. Most use the device's camera, but there are some that scan documents that are on your screen.

For resources in California Science Techbook, you'll need to sign in with your student's username and password the first time you access a QR code. After that, you won't need to sign in again, unless you log out or remain inactive for too long.

We encourage you to support your student in using California Science Techbook. May you and your student enjoy a fantastic year of science!

Sincerely,

The Discovery Education Science Team

Table of Contents

Unit 3: Life's Unity and Diversity

Unit Overview

Concept 3.1 Earth's History and the Fossil Record

Concept 3.2 Evolution and Natural Selection

Concept 3.3 Evidence for Evolution

Concept 3.4 Modifying Organisms

Unit Wrap-Up

Unit 4: Monitoring Biodiversity

Unit Overview

Concept 4.1 Nature of Waves

Concept 4.2 Waves and Matter

Concept 4.3 Warming Earth

Concept 4.4 Remote Sensing

Concept 4.5 Sustaining Biodiversity

Unit Wrap-Up

Resources

UNIT

3

Life's Unity and Diversity

Discovery EDUCATION

The End of the Dinosaurs

Dinosaurs, like *T. rex*, once ruled Earth. Why do they not exist today?

Quick Code
ca8501s

Guiding Questions

1. What can we infer about the history of life on Earth from the clues we can uncover in rock layers and the fossil record?
2. What evidence supports biological evolution?
3. How do evolution and natural selection explain life's unity and diversity?

Searching for a Cause

Solve Problems

Unit Project: Dinosaur De-extinction?

Quick Code
ca8502s

Thinking About Solutions

Could we bring back extinct animals?

You are part of a team of scientists. You are interested in an investigation to bring back extinct animals. All the animals will be kept in zoos. Before moving forward, determine the pros and cons to such an investigation. Consider what strategy you might use. Are there any ethical or moral problems you need to think about? Are there any technical difficulties you may encounter?

Dinosaurs Once Roamed Earth

Use the graphic organizer to sort your ideas.

For example, a "Pro" might be that it would add interest to zoos. A "Con" might be that the animals might not be easy to contain.

Pros

Cons

CONCEPT

3.1

Earth's History and the Fossil Record

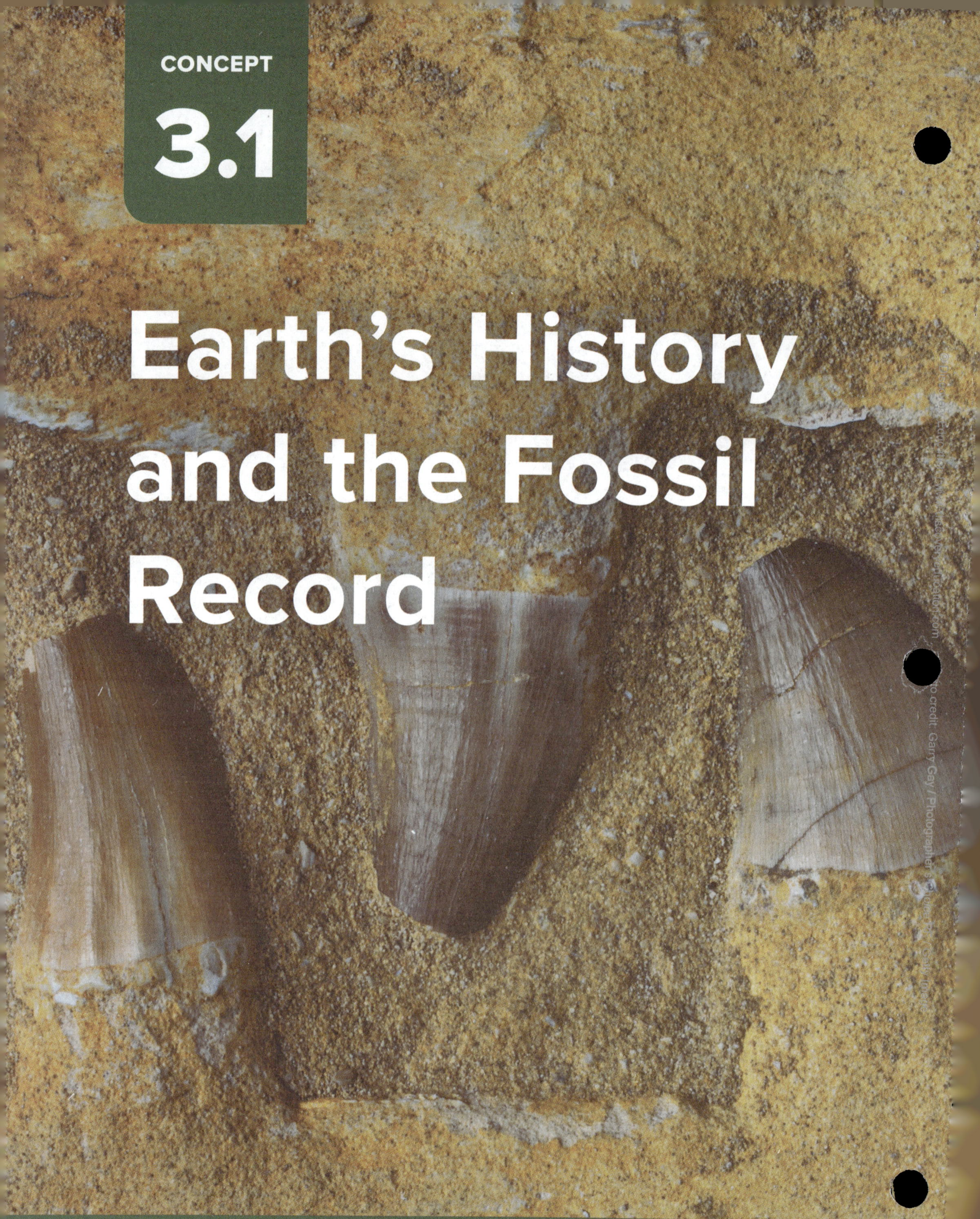

Student Objectives

By the end of this lesson:

- [ ] I can obtain, evaluate, and communicate information about how the formation of Earth led to conditions suitable for early life.
- [ ] I can use evidence from rock strata to construct a scientific explanation for how the geologic time scale is used to organize Earth's 4.6 billion-year history.
- [ ] I can synthesize information to construct explanations for how fossils form and why the fossil record is incomplete.
- [ ] I can analyze and interpret data from many sources to describe patterns in the fossil record and argue from evidence that fossils provide a record of past life and environments on Earth.
- [ ] I can analyze and interpret patterns in fossil evidence in rock strata to determine when organisms lived, how living things have changed over time, and what factors caused groups of organisms to flourish and decline.
- [ ] I can argue from evidence for the rate of evolutionary change and stability.
- [ ] I can use evidence from many sources to infer evolutionary relationships between modern and fossil organisms and to describe patterns of change and stability.

Key Vocabulary

air, amber, asteroid, atmosphere, bacteria, carbon, cast, decay, DNA, dwarf planet, fossil, gas, gravity, groundwater, hydrogen, liquid, magnetic field, mold, nebular hypothesis, nitrogen, outer core, oxygen, precipitation, sedimentary rock, solar system, solid, star, water, water cycle, water vapor, weather

Quick Code
ca8503s

Activity 1

Can You Explain?

What are fossils, and how effectively do they provide a record of the history of life on Earth?

Quick Code
ca8504s

Activity 2

Ask Questions

Changes in the Rocks

Quick Code
ca8505s

Like a mystery novel, our planet has hidden clues. Its clues are hidden not in pages but in layers of rocks. By looking at sequences of these rock layers and the clues within them, geologists can begin to reconstruct a story of Earth and solve some of its mysteries.

Video

Let's Investigate Changes in the Rocks

ENGAGE

As you watch the video online, **write** a list of questions that you would ask the scientists about the rocks and fossils found within them.

Activity 3

Analyze

Clues in the Rocks

Quick Code
ca8507s

Read the text. As you read, **highlight** text that identifies the different types of clues found in rocks that can help explain the origins of life on Earth. Then, **answer** the questions that follow.

Clues in the Rocks

If you have ever read a mystery novel, you will have noticed that, like most books, they have pages and chapters. Each page represents a small part of a story; each chapter pulls these parts together to build a bigger picture. Each chapter has clues that lead the reader closer to solving the mystery. The more mystery novels you read, the more skillful you become at solving the crime. Like a book, our planet has hidden clues. Its clues are hidden, not in pages, but in layers of rocks. By looking at sequences of these rock layers and the clues within them, geologists can begin to reconstruct a story of Earth and solve some of its mysteries.

What are some of the clues to be found in rocks? Some come from the types of rocks. Geologists can often tell the sort of environment in which a rock was formed—for example, how much oxygen was in the air at that time. They can learn this from the composition of the rock as well as the fossils that are found in it. Others come from the changes in rocks from one layer to the next. These layers can tell geologists about changes in Earth over long periods of time—millions or hundreds of million of years—or rapid changes.

CCC Patterns

What clues do rocks offer about how Earth was formed and the species that inhabit it?

How can geologists determine the environment in which a rock formed?

What other mass extinctions have you heard of in Earth's history?

What similar occurrences and/or events are being observed today that are affecting Earth's processes?

What questions do you have about fossils?

ENGAGE

Activity 4

Evaluate

Quick Code
ca8506s

What Do You Already Know About Earth's History and the Fossil Record?

Fossilization

This activity will test what you already know about the process of fossilization. **Read** the steps in the sequence of events below. Think about how a fossil may be formed. **Draw** a line to each step in the order you think a fossil will form.

organism is quickly covered in sediment	Step 1
organism's hard body parts fill with minerals	Step 2
organism dies and falls to the ground	Step 3
organism leaves an imprint of its shape	Step 4
organism decays over time	Step 5

ENGAGE

Important Geological Terms

Now, **check** your understanding of some key terms used in geology. **Write** a definition for each of the following terms.

Sedimentary Rock

Fossil

Radiometric Dating

How Did the Origin and Early History of Earth Enable Life to Originate?

Activity 5

Analyze

Quick Code
ca8509s

The Formation of Earth and the Characteristics of Earth That Enable It to Support Life

Choose one of the following sections of text: "The Formation of Earth" or "The Characteristics of Earth That Enable It to Support Life." **Read** the text and **watch** the videos.

If you selected "The Formation of Earth," you should **highlight** information that describes the steps of the nebular hypothesis. **On a separate sheet of paper, create** a cartoon with 5 to 8 segments illustrating the nebular hypothesis.

If you selected "The Characteristics of Earth That Enable it to Support Life," you should **highlight** evidence of Earth's characteristics that allow it to support life. **On a separate sheet of paper,** create a poster describing the three most compelling pieces of evidence from the text.

CCC **Cause and Effect**

CCC **Structure and Function**

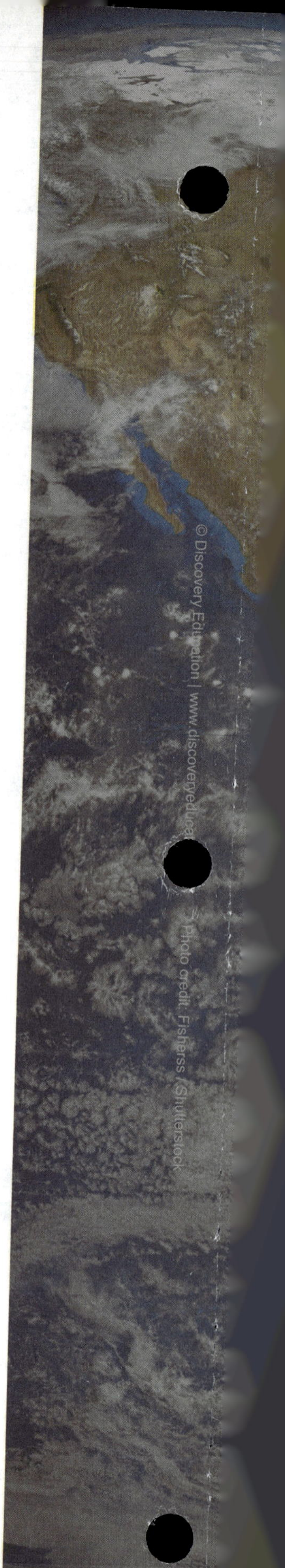

The Formation of Earth

EXPLORE

Earth formed as part of our solar system around 4.6 billion years ago. Scientists think that the solar system formed in a process that is described by the nebular hypothesis. According to the nebular hypothesis, the solar system formed from a giant rotating cloud of gas and dust, called a solar nebula. An unknown event, such as the shockwave caused by a supernova or the movement of a passing star, caused the cloud to begin to collapse under its own gravity, and the giant gas cloud began to contract. As the gas cloud collapsed, it took the shape of a flat disk of gas and dust, rotating faster and faster as it contracted. As material moved toward the center of the disk, the gravitational pull of the center increased, pulling in even more material. About 99.8 percent of all the material eventually formed a giant ball of gas and dust at the center of the disk. The pressure created by accumulating matter forced the material to heat up. The extremely high pressure and temperature at the center of the disk caused hydrogen atoms to fuse together to form helium. The star that we now know as the sun had formed.

Although most of the material that was originally in the nebula collapsed to form the sun, there was still a lot left over. This material continued to revolve around the center. Over time, it also came together under the force of gravity. Small rocks came together to form large rocks. These came together to form planetesimals, and planetesimals collided to form planets. Other clumps of material formed smaller objects such as the dwarf planets, moons, asteroids, and comets. This process is known as accretion.

Over time, as the collisions continued, some of the dwarf planets grew to form the eight planets of our solar system. One of the planets that was formed became Earth.

Early Earth

Cooling Earth

The Characteristics of Earth That Enable It to Support Life

EXPLORE

The position, size, and composition of Earth owe much to this early history. These, in turn, played an important role in the formation and evolution of life on our planet. Earth is unique in our solar system. It is the only planet known to support life. One reason that Earth can support life is its distance from the sun. Objects that are too close to the sun are exposed to too much solar energy, while those that are too far away do not get enough. Earth is at just the right distance for life to flourish. This region is called the habitable zone, meaning that it is the portion of our solar system where there is the right amount of solar energy for living things to survive.

Earth also has an atmosphere. Earth's atmosphere and distance from the sun help maintain an average surface **air** temperature of 15°C. Earth's atmosphere is made up of a mixture of gases, including **nitrogen, oxygen, carbon dioxide**, and **water vapor**. Most organisms need these gases to live. One reason Earth has an atmosphere is because of its mass. Earth is large and dense enough that its gravitational force is strong enough to hold the heavier gases, like nitrogen and oxygen, in its atmosphere. Lighter gases, like hydrogen and helium, can escape Earth's gravity, which is why they do not make up much of the atmosphere. Earth's atmosphere has a protective ozone layer, which shields life on Earth from much of the sun's harmful ultraviolet radiation. Earth's **magnetic field** also protects Earth's atmosphere from being stripped away by solar winds. Earth's magnetic field is the result of the convective motion of Earth's liquid metal in the **outer core**. Therefore, Earth's interior structure also helps it sustain life.

Earth's Magnetic Field

Water is essential to supporting life as we know it. Because Earth's temperatures are moderate, water can exist in all three states in Earth's system—as a solid, liquid, or gas. The oceans are made of liquid water, and the atmosphere is partially composed of water vapor, which is water in gas form. The atmosphere is also where Earth's **weather** takes place, which contributes to the movement of water and air around the planet.

Water on Earth

Water is a key factor for life on Earth because all living things require water. The **water cycle** involves the movement of all different forms of water, including water vapor, **precipitation**, water in rivers and streams, ocean water, **groundwater**, and ice.

Activity 6

Reason

Quick Code
ca8510s

Geologic Time

In this investigation, you will create a model of Earth's history and use it to gain a perspective on the huge scale of time it covers.

What materials do you need? (per group)

- Measuring tape
- Calculator
- Copy of the geologic time scale

EXPLORE

SEP Analyzing and Interpreting Data

SEP Developing and Using Models

SEP Using Mathematics and Computational Thinking

Procedure

Read the geologic time scale, and examine the table listing the intervals.

Eon	Era	Period	Beginning (Millions of Years Ago)
Phanerozoic	Cenozoic	Quaternary	2.6
		Neogene	23
		Paleogene	66
	Mesozoic	Cretaceous	145
		Jurassic	201
		Triassic	252
	Paleozoic	Permian	299
		Carboniferous (Pennsylvanian)	323
		Carboniferous (Mississippian)	359
		Devonian	419
		Silurian	444
		Ordovician	485
		Cambrian	541
Proterozoic			2,500
Archean			4,000
Hadean			4,550

Note both the names of geologic eras and the time range in millions of years for each one. What ways can you think of to represent this time scale in a model?

Next, use the tape measure and calculator to determine a reasonable scale for your model. Then, fill out the table below with your measurements.

Geologic Period	Duration (Millions of Years)	Length (Yards or Meters)
Quaternary	2.6	
Neogene	20.4	
Paleogene	43	
Cretaceous	79	
Jurassic	56	
Triassic	51	
Permian	47	
Pennsylvanian	24	
Mississippian	36	
Devonian	60	
Silurian	25	
Ordovician	41	
Cambrian	56	

Go to a space that fits your scale (a hallway or large classroom). Then, choose students to represent the start of each geologic period and have them move to the correct position in the model based on your table.

What does this model show you about the scale of geologic time?

____

____

How would the model have been different if you had used geologic eras instead of periods?

____

____

EXPLORE

Reflect

How do scientists determine where a period ends and a new one begins?

What are some of the organisms or events that happened in Earth's history, based on the geologic time scale?

How Do Fossils Record the Past of Life-Forms?

Activity 7

Evaluate

Quick Code
ca8511s

EXPLORE

The Oldest Fossils

Look at the image of layers of rock below Earth's surface, and **think** about how this layering develops. Which rocks are older and which are newer? **Look** at the fossils and **read** their age in millions of years. **Write** each fossil in the appropriate layer of rock. Pay close attention to where you write it. Should it go near the top of a layer, the middle, or the bottom?

- human bone (10,000 years old)
- fish (200 million years old)
- dinosaur bone (90 million years old)
- horseshoe crab (220 million years old)
- fern leaf (50 million years old)

Activity 8
Analyze

Fossils Record the Past

As you read the text, **write** an explanation of how fossils show evidence of once-living organisms. Be sure to support your argument with evidence from the text. Then, **share** your explanation with the class.

Quick Code
ca8512s

Fossils Record the Past

Fossils are the preserved evidence of once-living organisms. Some fossils come from the remains of an organism, such as shells, bones, or teeth. These are called body fossils. Other fossils come from the traces of an organism's activities, such as their feces, footprints, or burrows. These are called trace fossils. Fossils exist from a wide variety of organisms—from microscopic **bacteria** to giant trees and dinosaurs. They can be found in many places on Earth, but almost all are preserved in **sedimentary rocks**.

Sedimentary rocks are formed as sediment, such as silt or clay, and settle in layers. Quickly settling sediment is an ideal environment for **fossil** formation and preservation. The faster the burial, the more likely it is for a fossil to form. This is because rapid burial protects dead organisms from **decay** and scavengers. If sediment covers an organism's remains before decay is complete, parts of the organism can be preserved. Over time, the organism may become fossilized.

SEP Engaging in Argument from Evidence

This can happen in a variety of ways. A whole organism can be preserved in ice, **amber**, or tar. An organism can decay and leave a detailed imprint of itself, called a **mold**, inside the rock. The mold can then be filled with minerals to form a copy of the organism called a **cast**. In a carbonized fossil, the **carbon** in an organism's body leaves a thin film in rock that preserves the shape of the organism. With replacement fossils, a bone or shell is partly or completely replaced with minerals so that only some of the original material remains—or none at all—but the shape is identical.

EXPLORE

An organism's body does not have to be fossilized to leave evidence of its existence. Trace fossils are not made from the bodies of organisms but by their activities. Like molds, trace fossils are made from impressions in sediment. However, instead of an impression of an organism's body, trace fossils are impressions of footprints, burrows, and other shapes left by the movement of organisms.

Fossils in Amber

Fossils are rarely complete and can never provide a complete record of what an organism looked like. Some information is always lost. One reason is that many parts of an organism decay too quickly to become fossilized. Most fossils are evidence of the hard parts of organisms, such as bones, teeth, or shells.

Fossils Record the Past *cont'd*

The soft parts, such as organs, are rarely preserved. For example, fossils of snail shells are common, whereas fossils of slugs, which are like snails but lack shells, are rare. Additionally, it is very rare for any organism to become fossilized, and therefore the fossil record does not provide a complete picture of life in the past.

Next, **write** down one comment and one question on sticky notes about each student's explanation, and **pass** them to the student after their presentation. **Review** the comments and questions you received from other students, and **write** a response to one of the questions.

EXPLORE

Activity 9

Evaluate

What Becomes a Fossil?

Quick Code ca8513s

Examine the organisms in the three pictures below. If these organisms were all buried under similar conditions, how much of their bodies would be preserved as fossils?

Think about which organism would be most preserved if it became a fossil. **Draw** lines to put them in order from least preserved to most preserved.

Least		Most

How Do Scientist Interpret the Fossil Record?

Activity 10

Analyze

Interpreting the Fossil Record

Quick Code
ca8514s

As you read the text, **highlight** information about relative and absolute dating.

EXPLORE

Interpreting the Fossil Record

The fossil record helps scientists interpret the story of life on Earth. However, inevitably, some parts of the story are missing. Think of the fossil record as being like a book with some words or pages missing. We have a good idea how the story unfolds, but some details or clues have yet to be revealed. Despite these missing pieces, the fossil record provides an excellent picture of how life and Earth's environments have changed through time.

Because rocks are laid down in sequence, with the oldest rocks first, scientists can place the organisms that formed in them in chronological order—usually the oldest at the bottom and the youngest at the top. Scientists may not know the age of the rock layers and the fossils in them, but they can determine their ages compared to one another. This is called relative dating. Luckily, scientists have another tool at their disposal—absolute dating. Absolute dating, although not exact, uses radioactive elements in rocks and gives a rock's age as a range.

Interpreting the Fossil Record *cont'd*

For this reason, it is also called radiometric dating. Absolute dating works most effectively on igneous rocks. If a sequence of rocks has an igneous layer, such as a lava flow, the rocks above and below it can usually be dated. By combining information from absolute dating at different sites around the world, with the types of fossils in adjacent rocks, scientists have pieced together a fairly accurate "calendar" of Earth's history. This calendar is known as the geologic time scale. If a fossil species is widespread and indicative of a particular time in Earth's history, it can be used to identify the age of a rock. Such fossils are called index fossils.

Next, **compare** what you highlighted with that of another student. **Share** your thoughts on relative and absolute dating, and **answer** the following questions.

What information can absolute dating give a scientist about the fossil record?

EXPLORE

What information can relative dating give a scientist about the fossil record?

Why is it important for scientists to consider both absolute dating and relative dating?

Activity 11

Reason

Quick Code
ca8515s

Relative Dating

In this investigation, you will look at a group of fossil samples and determine which are the oldest and which are the youngest.

What materials do you need? (per group)

- Fossil sample cards, 9

SEP Analyzing and Interpreting Data

Procedure

Review the nine fossil sample cards with your group. Each sample was collected from a different rock layer in a geologic column. Sample 1 came from the lowest layer in the column, but the other samples have become mixed up and are out of order. Work with your team to arrange the samples from oldest to youngest.

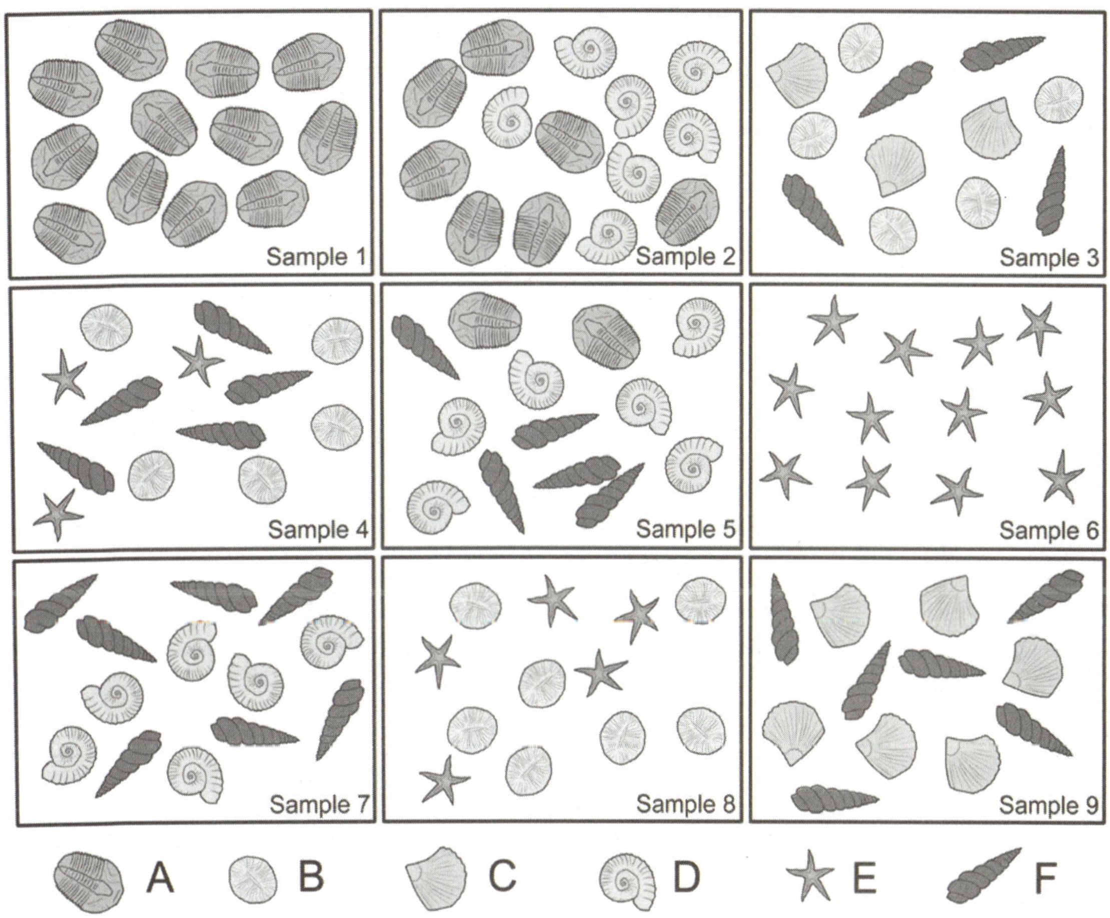

EXPLORE

How did your group decide the age of each fossil sample?

Reflect

In what order did you arrange your fossils?

Which fossil is the youngest? Explain your reasoning.

Which organism lived for the longest period of time? Which lived for the shortest period of time?

In which samples was biodiversity the greatest? In which samples was biodiversity the lowest? Explain your reasoning.

EXPLORE

What Are Some of the Patterns That Exist in the Fossil Record?

Activity 12

Observe

Quick Code ca8516s

The Permian Extinction

The process of dating fossils reveals patterns in the fossil record. One of the most obvious is that fossil species or groups of species occupy certain time intervals in geologic history. Most stay in the fossil record for a period of time and then become extinct. In fact, one of the most immediate observations we can make when we look at the tree of life is that organisms that existed in the past are different from those that are around today.

Watch the video online, and then **answer** the questions.

The Permian Extinction

What is a mass extinction?

What living things began to increase in numbers following this mass extinction?

What other mass extinctions have you heard of in Earth's history?

EXPLORE

Activity 13

Analyze

Quick Code
ca8517s

Patterns in the Fossil Record

When we look at the fossil record, we can discern certain patterns. These include species becoming extinct, mass extinctions, adaptive radiation, gradual versus rapid change, and relationships between species and transitional forms.

Read the text about each of these patterns. **Highlight** main ideas and key details as you read. Then, pair up with another student and **fill out** the chart together.

CCC Cause and Effect

Patterns in the Fossil Record

Species Become Extinct

The process of dating fossils reveals patterns in the fossil record. One of the most obvious is that fossil species or groups of species occupy certain time intervals in geologic history. Most stay in the fossil record for a period of time and then become extinct. In fact, one of the most immediate observations we can make when we look at the tree of life is that organisms that existed in the past are different from those that are around today. Species, in fact whole groups of species, have appeared, diversified, and then disappeared. All species eventually become extinct. Over the past 3.5 billion years, about 10 billion species have existed. Of those, only around 5 to 16 million are around today. The reason for this extinction varies. It may be competition from other species or changes to the environment that cause a specific organism to become extinct. Left undisturbed, the tree of life increasingly branches, as organisms diversify into many different forms. Then, suddenly, the number of organisms decreases as huge numbers of species go extinct at around the same time. When this happens on a large scale, it is called a mass extinction.

Mass Extinctions

A pattern of mass extinctions is observable in the fossil record. The table summarizes the past five that have been identified, and one that is occurring now.

Major Extinction Events

Extinction Events	Time Millions of Years Ago	Impact on Species
Ordovician-Silurian	440–450	27% of families, 57% of genera
Late Devonian	360–375	19% of families, 50% of genera, 70% of species
Permian-Triassic	251	57% of families, 80% of genera, 96% of marine species, 70% of land species
Triassic-Jurassic	205	23% of families, 48% of genera
Cretaceous-Tertiary	65.5	17% of families, 50% of genera, 75% of species
Quaternary	NOW	ongoing and accelerating...

The causes of these past mass extinctions are not fully understood, and evidence points to a variety of different events. These events include asteroid impacts, huge volcanic eruptions, massive releases of methane gas, and climate change. The largest mass extinction occurred around 250 million years ago. In this extinction, nicknamed the "Great Dying," about 96 percent of species were wiped out. Its most likely cause was a huge release of lava accompanied by other changes. Perhaps the most widely understood mass extinction occurred around 65–66 million years ago. This was most likely caused by the impact of an asteroid and has been held responsible for the disappearance of the dinosaurs.

Adaptive Radiation

After mass extinctions, many of the niches that species once occupied are left vacant, and new species arise to replace them. This sudden rise in species diversity after a mass extinction is another pattern that can be observed in the fossil record and is called adaptive radiation. Adaptive radiation occurs when a trait evolves that enables a group to survive in a new habitat. Adaptive radiation also occurs when a trait enables new life-forms to outcompete old ones. One example of this is the rapid diversification of flowering plants during the Cretaceous that displaced many giant ferns and conifers.

Relationships between Species and Transitional Forms

Each new fossil species that is discovered represents a snapshot in the evolution of that species over time. Scientists can use the fact that fossil species with similar features are usually related to build family trees or evolutionary trees for closely related species. You can think of these as twigs on the branches of the larger tree of life. On these trees, the joints between branches represent common ancestors.

An Example of a Simple Evolutionary Tree

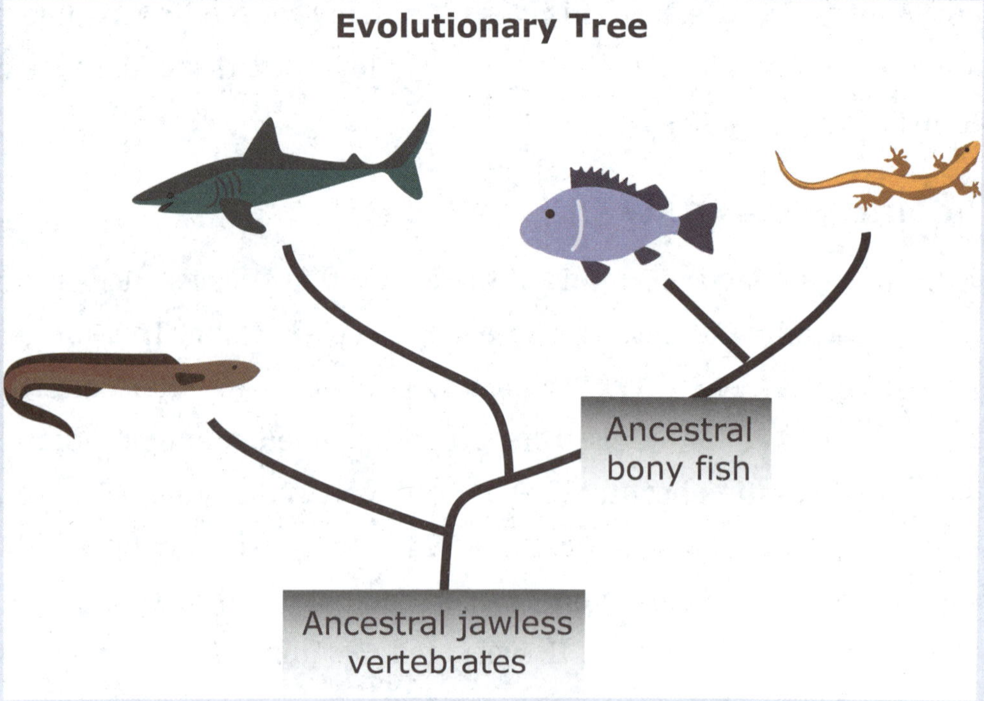

This means that evolutionary trees can be used to predict the existence of these common ancestors, and if the scientists know their age and the types of rocks in which they are found, scientists can pinpoint their searches for these transitional forms—organisms that show traits of both. However, when they find them, scientists can never be sure whether they are transitional forms or other branches of the tree. Scientists usually combine fossil information with other information to create evolutionary trees. You will learn more about this later.

With a partner, **complete** the chart.

Phenomenon	Describe What a Scientist Examining the Fossil Record Would See to Determine the Phenomenon
A species going extinct	
A mass extinction	
Adaptive radiation	

EXPLORE

Activity 14

Observe

New York Ice Age

Quick Code
ca8518s

When we examine the fossil record, we can observe similarities between ancient organisms and modern ones.

As you watch the video, **see** if you can **identify** modern animals that are similar to the ancient animals that once lived in what is now New York City.

Video

New York Ice Age

CCC Structure and Function

Use the graphic organizer to **compare** ancient animals from the video to modern animals.

Ice Age Animal	Modern Animal

EXPLORE

What Can the Fossil Record Tell Us about the Ecology and Behavior of Fossil Organisms?

Activity 15

Investigate

Hands-On Investigation: Fossil Tracks

Quick Code
ca8519s

In this investigation, you will examine fossil images to discover how an animal moved when it was alive.

Predict

What can trace fossils, such as footprints, burrows, and trails, tell us about how an animal lived?

CCC Cause and Effect

SEP Analyzing and Interpreting Data

SEP Using Mathematics and Computational Thinking

What materials do you need? (per group)

- Paper
- Pencils
- Measuring tape
- Sand, fine
- Images provided by teacher

HANDS-ON INVESTIGATION

EXPLORE

Procedure

1. Divide into pairs. Measure the distance from your hip to the floor in centimeters, starting from the ilium, or the uppermost, largest part of the hip bone. Then, measure your foot, also in centimeters. Record your measurements using the following ratio:

Fl:h

Fl = foot length

h = distance from hip to floor

What do you notice about the ratio?

2. Now, you and your partner will go to the sandbox or sandy area and rake it until it is smooth and level. Then, walk across the sand and measure the distance between your steps, measuring from one foot (left or right) to the same foot. Do this several times, and then calculate the average. This is your stride length (SL). Write your average SL in the space below.

3. Rake the sand smooth again, and then run across the sand. Once again, measure the distance between your steps, measuring from one foot (left or right) to the same foot. Do this several times, and then calculate the average. This is your running stride length (RSL). Write your average RSL in the space below.

4. Rake the sand smooth again, and then jog or trot across the sand (this should be slower than the running). Once again, measure the distance between your steps, measuring from one foot (left or right) to the same foot. Do this several times, and then calculate the average. This is your trotting stride length (TSL). Write your average TSL in the space below.

How can this information help scientists who study fossils of animal tracks?

5. Now, **look** at the three images of animal footprints.

What inferences can you make about the behavior of the animals that made these tracks?

EXPLORE

Reflect

What are some of the characteristics of the animals that left these tracks?

What are the similarities and differences among the animals that left these tracks?

EXPLORE

Activity 16

Analyze

What the Fossil Record Can Tell Us

Quick Code
ca8520s

Read the text and **highlight** evidence to answer the question: What can the fossil record tell us about the ecology and behavior of fossil organisms?

What the Fossil Record Can Tell Us

Fossils can provide information on the structure of now-extinct organisms. The structures in fossils can be compared with those in modern organisms. From these structures, aspects of their behavior can be inferred. For example, scientists think that many long-necked dinosaurs ate leaves from high in trees, much like modern giraffes. Other structures indicate how organisms may have defended themselves from predators or displayed themselves to potential mates.

Scientists can use other information to find out about ancient diets. For example, examination of animal teeth can reveal microscopic wear—the marks left by the food an animal eats. A study of the teeth of our four-million-year-old ancestor, best known by the famous fossil of Lucy, shows she ate tough grasses and leaves. Some fossils have fossilized food in their stomachs. These stomach contents can shed light on what they once ate—one fossil mammal was found to have had a tiny dinosaur for its last supper. Fossil feces are also often found. Analysis of this fossilized bodily waste provides information on diet. Such evidence reveals the feeding relationships between organisms in ancient ecosystems.

CCC Cause and Effect

These ancient ecosystems can be better understood when different fossils are found and studied together. Imagine if a modern coral reef was suddenly covered with mud and eventually became fossilized. What fossils would you find a million years later? What information could you glean from how the organisms were positioned in the reef? You would not only have a complete fossil ecosystem, you would be able to infer the environment in which it existed, in this case a warm saltwater sea. Scientists can interpret real groups of fossils in the same way.

Changes in the distribution of fossils or the species found in a certain area can be used to infer changes in climate. Additionally, fossils of the same organism found in widely separate areas can tell scientists a lot about the ancient positions of continents.

Trace fossils, such as footprints, can be used to work out how animals moved and interacted. The length of stride and shape of the prints can provide information on how fast an animal, such as a dinosaur, was moving. Groups of tracks of the same species could be an indication that the animals were social—perhaps living in herds or packs. The size of the prints provides an indication of the size of the animals as well as, perhaps, the numbers of adults and young.

Burrows and nests can tell us how animals lived and raised their young. For example, a study of about 30 different dinosaur nesting sites suggests that many species buried their eggs (as do modern crocodiles), whereas others have been found in open nests similar to those of modern birds.

With a small group, **create** a list of things the fossil record can tell us about the ecology and behavior of fossil organisms.

EXPLORE

Activity 17

Record Evidence

Changes in the Rocks

Quick Code
ca8521s

As you worked through this lesson, you investigated and gathered evidence about Earth's history and the fossil record. Now, take another look at the video Changes in the Rocks, which you first saw in Engage.

Let's Investigate Changes in the Rocks

Can You Explain?

What are fossils, and how effectively do they provide a record of the history of life on Earth?

Use your new understanding of Changes in the Rocks to write a scientific explanation answering a question.

1. Choose a question. You may choose to answer the Can You Explain? question or one of the questions you wrote at the start of this lesson.
2. Use the following information to help you write your explanation.

SEP **Constructing Explanations and Designing Solutions**

You should use a scientific explanation to answer this question. Recall that a scientific explanation contains three elements: a scientific claim, evidence to support the claim, and reasoning that connects the evidence to the claim. As you plan your scientific explanation, be sure to include:

- reliable evidence from sources, including readings and/or videos
- data that provide evidence from investigations
- references to theories and/or laws that describe the natural world
- reasoning that shows how the evidence supports the claim

Record your notes in the spaces provided.

Using your notes, write your scientific explanation for the Can You Explain? question in the spaces provided here.

Write the Can You Explain? question in this space.

Scientific Explanation (Explanation = Claim + Evidence + Logical Reasoning)

Directions:	Key Components:
Create a scientific explanation for the question. This combines the claim that was made, the evidence that was collected, and a justification for why the evidence supports the claim (logical reasoning). The explanation should contain clear thoughts and accurate scientific vocabulary.	• Uses precise and accurate language • Uses scientific vocabulary • Provides clear, logical thoughts • Uses evidence and reasoning to support the claim

EXPLAIN

STEM in Action

Activity 18
Analyze

Living in the Past: Careers in Earth History

Quick Code
ca8522s

Read the following text, and **view** the videos.

Living in the Past: Careers in Earth History

Fossils provide some of the best evidence for the existence of extinct animals in Earth's history. They also provide evidence for the evolution of different organisms over time. Fossils are all that remain of ancient organisms to tell us what they looked like and how they lived. You may have seen dinosaurs or other fossils in a museum, but did you ever wonder where they came from? Obtaining fossils isn't as easy as just digging in the dirt with a shovel. Fossils are made of rock, but they can be very fragile. They have to be excavated very carefully to avoid damaging them. Let's take a closer look at how paleontologists excavate the fossils they study and display in museums.

Fossils are not the only tools scientists can use to study changes in Earth's history. The fossil record tells the story of ancient life on Earth. The rock record tells the story of changes in Earth itself.

SEP Obtaining, Evaluating, and Communicating Information

SEP Engaging in Argument from Evidence

Digging for Fossils

Climate change is an important topic today. We wonder how we can address the problems of climate change and what the future holds. Some scientists, however, study climate change of the past rather than the future. By studying the geochemistry and geology of the rock record, and combining that information with the fossil record, scientists can understand how Earth's climate has changed in the past. Scientists study a period of rapid global warming that occurred 55 million years ago.

Unearthing Ancient Climate Change

ELABORATE

Living in the Past: Careers in Earth History *cont'd*

In addition to the fossil and rock records, scientists can study ice in Earth's polar regions to learn about climate change. Scientists who study climate change in Earth's past are called paleoclimatologists. Paleoclimatologists can use the chemistry of the ice to learn about the history of Earth's climate. They can also use the chemistry of gas bubbles trapped in the ice to learn about the history of Earth's atmosphere. This ice is a valuable record of Earth's past. Unfortunately, this record is disappearing due to the climate change that is happening now. The global warming trend is causing the polar ice caps and glaciers to melt, destroying the valuable ice record. However, scientists can use the rate of this melting to learn about present-day climate change. Scientists also study both historical and modern climate change.

Can you think of other careers that need to have knowledge of changes in Earth's history?

Video

Archived in Ice: Rescuing the Climate Record

Evidence of Evolution

In this interactive, you will investigate how scientists piece together the events of Earth's history. First, you will interpret the relative age of a geologic section of sedimentary rock layers. Then, you will determine the absolute age of those layers by analyzing the fossils they contain.

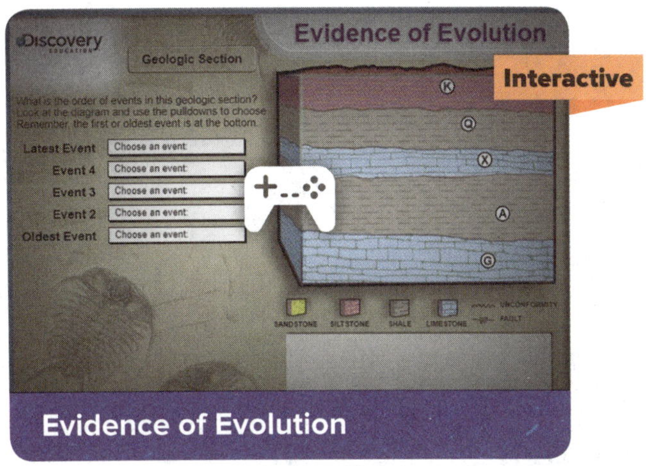

Evidence of Evolution

Relative Age vs. Absolute Age

How is relative age different from the absolute age of a geologic or evolutionary event?

____

____

____

____

ELABORATE

How Do They Know?

How can scientists know what happened millions of years ago if no one was there to see it?

The Fossil Record

How does the fossil record provide evidence that life has changed over time?

Determine the Age of Shale

Suppose you find a fossil in a layer of shale. How might you determine the age of the shale?

Differences between Fossils

How might the types of fossils found in a million-year-old layer of rock differ from fossils found in a 200-million-year-old layer?

ELABORATE

Activity 19

Concept Review

Review: Earth's History and the Fossil Record

Quick Code
ca8523s

Now that you have completed the objectives for this concept, **review** the core ideas you have learned. Record some of the core ideas below.

Core Ideas

EVALUATE

Talk with a Group

Now, think about the video Searching for a Cause that you saw in Get Started. Discuss how what you've learned about Earth's history and the fossil record can help you understand the search for the cause of dinosaur extinction.

CONCEPT

3.2

Evolution and Natural Selection

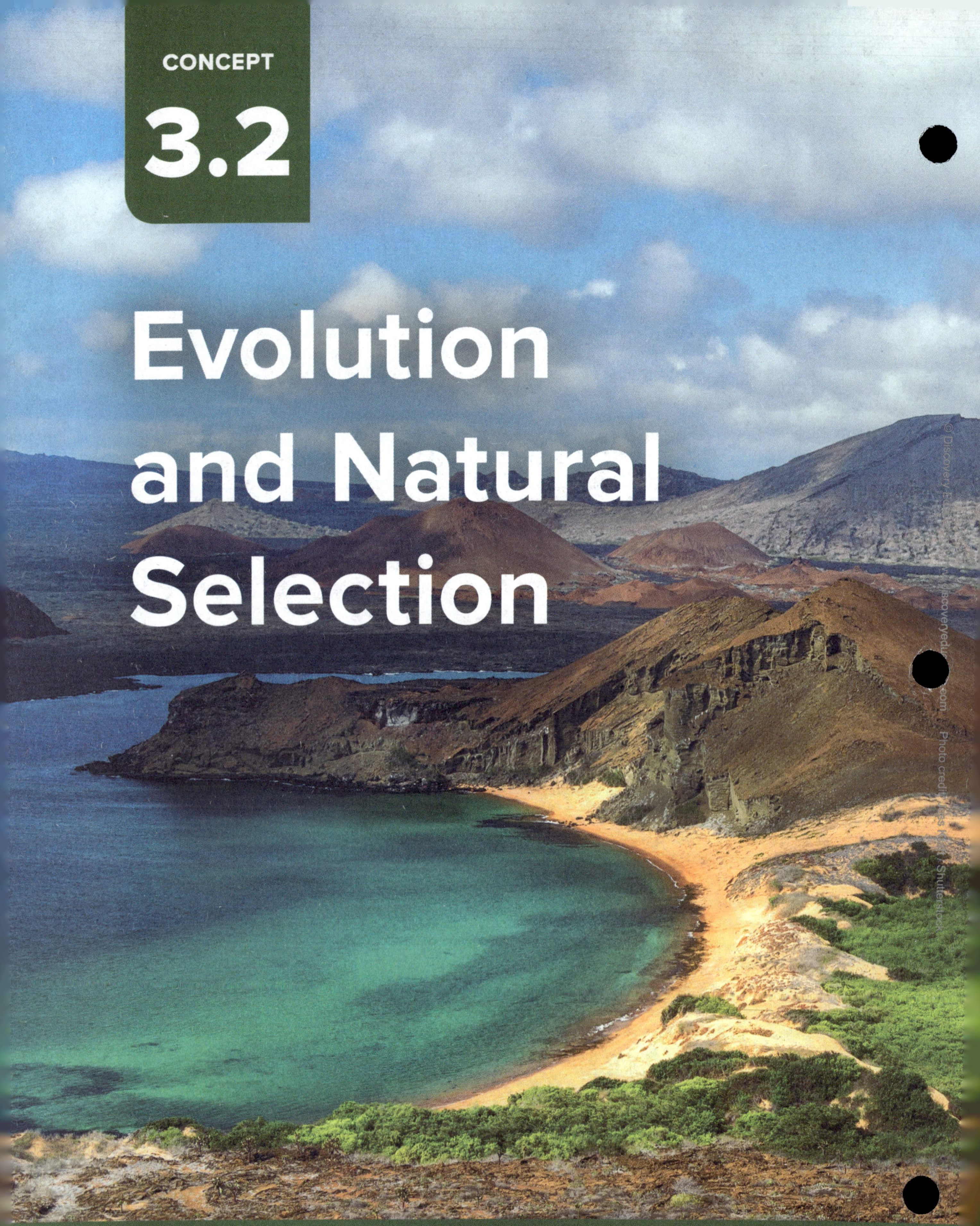

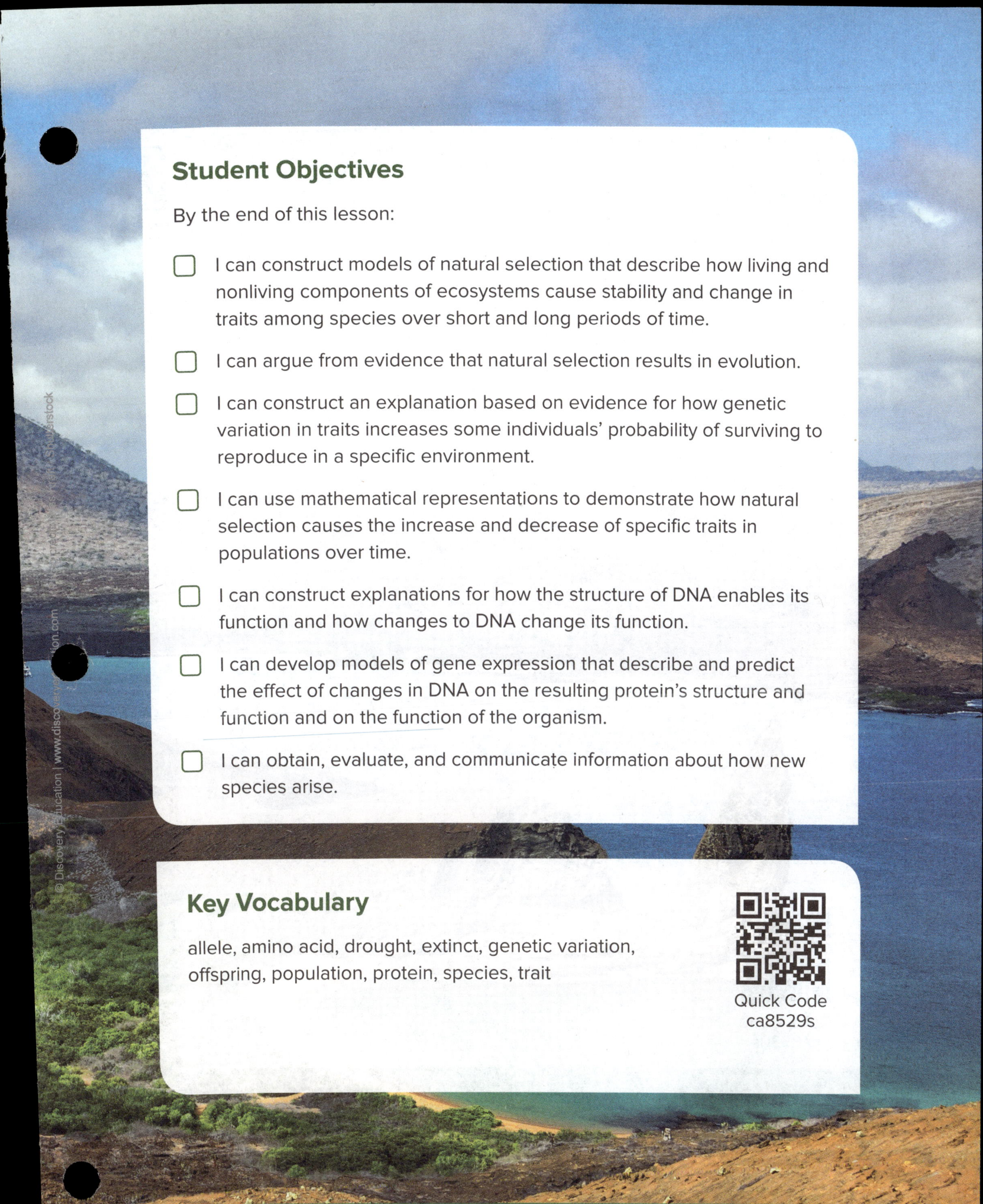

Student Objectives

By the end of this lesson:

- [ ] I can construct models of natural selection that describe how living and nonliving components of ecosystems cause stability and change in traits among species over short and long periods of time.
- [ ] I can argue from evidence that natural selection results in evolution.
- [ ] I can construct an explanation based on evidence for how genetic variation in traits increases some individuals' probability of surviving to reproduce in a specific environment.
- [ ] I can use mathematical representations to demonstrate how natural selection causes the increase and decrease of specific traits in populations over time.
- [ ] I can construct explanations for how the structure of DNA enables its function and how changes to DNA change its function.
- [ ] I can develop models of gene expression that describe and predict the effect of changes in DNA on the resulting protein's structure and function and on the function of the organism.
- [ ] I can obtain, evaluate, and communicate information about how new species arise.

Key Vocabulary

allele, amino acid, drought, extinct, genetic variation, offspring, population, protein, species, trait

Quick Code
ca8529s

Activity 1

Can You Explain?

What is natural selection, and how does it work?

Quick Code
ca8530s

Activity 2

Ask Questions

The Medium Ground Finch

Quick Code
ca8531s

Let's Investigate the Medium Ground Finch

Consider with Your Group

This picture came from a book written by the famous Victorian naturalist Charles Darwin. What do you notice in this image?

What questions do you have about this bird?

ENGAGE

Activity 3

Analyze

Finches and Famine

Quick Code
ca8549s

Read the following text, and **answer** the questions that follow.

Finches and Famine

Imagine spending years studying birds on a remote island in the Pacific Ocean. This is what two scientists have done on the deserted island of Daphne Major, one of 31 islands that make up the Galápagos Islands, which lie on the equator in the Pacific Ocean.

This remote island turned out to be a natural laboratory for studying changes in bird populations in response to changes in the environment. Daphne Major is an island of extremes, with periods of long, severe **drought**, followed by periods of extended rainfall. With these extreme changes come changes in the food that the birds can obtain—feast and famine. The scientists, Mary and Peter Grant, who originally intended a study lasting two years, studied these birds for almost 30 years. Why did they study these birds for such a long time?

The birds they were studying were members of 13 closely related **species** of birds known as Darwin's finches. These bird species differ in the types of food they eat. In particular, their beaks differ in size and shape depending on their diet. These differences occur not only between species but within species. The Grants' research focused on one of these species—a rather dull-looking bird called the medium ground finch, *Geospiza fortis.*

CCC Stability and Change

ENGAGE

Normally, medium ground finches use their small beaks to eat small seeds. In times of drought, these tiny seeds become scarce. Only big seeds are available, and only those medium ground finches with slightly bigger beaks can eat them. The Grants found that a year after a drought, the average beak size in the **population** had increased. The newest generation of finches had beaks around 4 percent bigger than their grandparents. How and why do you think this change had taken place?

What the Grants had observed was an example of evolution by natural selection. You will learn more about these two terms and how they are closely linked as you explore this concept.

On what did the Grants focus their research in the Galápagos Islands? What observations did they make?

What is the mechanism behind these changes? What is required for these changes to occur?

Activity 4

Interpret Data

Finch Data

Quick Code
ca8532s

Use the graphs to help answer the questions below.

Change in Seed Size
(1976–1978)

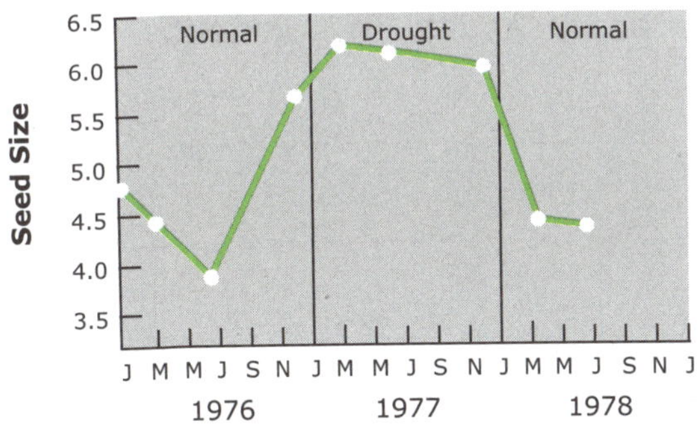

Distribution of Bill Depth and Survival Rate in Medium Ground Finches

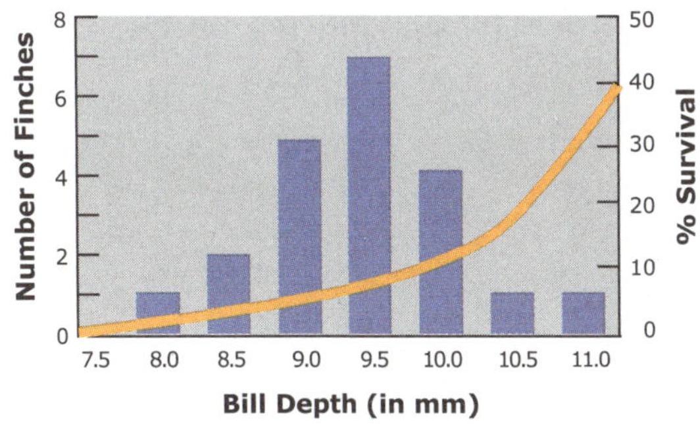

What patterns do you observe in the data presented in the graphs?

____

____

____

How could this have possibly resulted in changes in later generations of medium ground finches?

____

____

____

CCC **Stability and Change**

Activity 5
Evaluate

Quick Code
ca8533s

What Do You Already Know About Evolution and Natural Selection?

Evolution Basics

In your group, **choose** an animal in a specific environment. For example, consider a squirrel in an oak-hickory forest. **Describe** your animal and its environment.

Next, **propose** an environmental challenge to which the animal must respond through evolutionary change. Examples could include reduced habitat due to flooding or competition for food due to a new invasive species in the area.

ENGAGE

Exchange challenges with other groups. How will the animal evolve based on adaptations to environment change? **Explain** your reasoning.

____

____

____

____

____

____

____

____

____

____

____

____

____

Which statements correctly describe factors that are necessary for evolution to occur? **Select** all that apply.

- ☐ Genetic variation in a population must be present.
- ☐ Individuals within a population must be able to reproduce.
- ☐ Resources in an environment must be abundant.
- ☐ Variation that does not affect the survival of a species must be present.
- ☐ A stable physical environment that does not change must be present.

How Is Individual Variation in a Population Useful?

A population of animals lives in an environment with plenty of low-hanging fruit to eat. Over time, the environment changes. An extended drought causes food to become scarce. Most of the low-hanging fruit is quickly eaten, but fruit higher up on trees is still available to eat. Which animal would be most likely to survive and reproduce in this new environment?

- ◯ an animal with a very flexible neck
- ◯ an animal with a very short neck
- ◯ an animal with a very long neck
- ◯ an animal with a very thin neck

Come up with a new scenario in which the environment changes in a new way. **Exchange** scenarios with a partner, and **identify** an animal or plant that could survive in this new situation.

Photo credit: Jess Kraft / Shutterstock

What Was Darwin's Big Idea?

Activity 6

Analyze

Darwin's Big Idea

Quick Code
ca8534s

As you read the following text, **highlight** information that answers the following questions:

- Who was Charles Darwin?
- What did Charles Darwin observe about the organisms on the isolated Galápagos Islands?

Darwin's Big Idea

Nobody is sure how many **species** of organisms exist on Earth. Estimates range between 8 and 15 million. Some scientists think there are even more than that. We know from the fossil record that organisms have appeared and then become **extinct** during Earth's history and that today's species are just a small fraction of all those that have existed. We can also see relationships between fossil species and those that are alive today. The fossil evidence suggests that modern species are related to, and derived from, previously extinct forms. How does this happen? How can one type of animal or plant become another? Where do species come from?

These questions were similar to ones asked by a young naturalist who, in 1831, embarked upon a five-year voyage of discovery on the

CCC **Stability and Change**

British survey ship HMS *Beagle*. The young naturalist's name was Charles Darwin. He was an avid collector of plants, animals, and their fossils. During his voyage, he collected thousands of specimens—many of species new to science. He was amazed by the huge diversity of organisms and began to wonder how so many species came to be. He also studied their distribution, particularly on a small group of islands that lie off the coast of Ecuador—the Galápagos Islands. In the Galápagos Islands, he wondered why species of birds and tortoises were different on islands that were only a few miles apart. When he returned to England, he began to study his collections, organize his field notes, and develop hypotheses for some of his questions.

In 1859, after many years of painstaking research, Charles Darwin published *On the Origin of Species*. The book contained a big idea that would change the world: the hypothesis of evolution by natural selection. It explained where different species came from. At the time, most of his evidence came from the fossil record and his observations of living creatures. Now, his hypothesis is supported by more than a century of scientific studies and is considered a scientific theory. What exactly is the theory of evolution by natural selection? You may have heard of it by another name: survival of the fittest. How does it work?

EXPLORE

Activity 7

Analyze

Darwin and Artificial Selection

Quick Code
ca8535s

As you read the following text, **highlight** evidence that helps explain how the driving forces behind natural selection and artificial selection differ.

Darwin and Artificial Selection

Charles Darwin and finches will forever be linked in history. Darwin's book *On the Origin of Species* featured these birds. But did you know that Darwin studied another bird, the pigeon, to help him support his idea about why there are many different varieties of plants and animals on Earth?

During and even long before Darwin's time, it was fashionable to breed pigeons. The pigeons were used as message carriers and as pets. They were even used in races!

Darwin noticed that the reason for owning pigeons influenced the characteristics the pigeon breeders crossed. For example, breeders who had pet pigeons bred pigeons that had big chests. Those who used pigeons for racing crossed pigeons that were fast so they could produce faster **offspring**. After many generations of breeding, the offspring pigeons had physical, internal, and behavioral characteristics that were very different from the original parents.

Darwin did not only depend on what he observed from the breeders' pigeons though. He became a breeder himself, and he did his own experiments. The results of his experiments led him to believe that all the different pigeons he studied came from the wild rock pigeon. He also observed that there were many differences between the pigeons that humans bred and the pigeons that lived in the wild.

Darwin then realized that he could use his observations to support his theory of natural selection. Just as humans artificially selected which individual birds to breed, Darwin saw that the environment imposed a natural selection of individuals that would breed. In natural settings, only the individuals that were well suited to their environment survived long enough to reproduce. For example, the predecessors of domestic pigeons were wild rock pigeons. These birds lived in an environment where there was stiff competition for food and mates, as well as predators at every turn. Only the individuals with traits that enabled them to outcompete other individuals for food and mates and evade predators would be able to survive to adulthood, when they could produce offspring. Thus, Darwin was able to link the concepts of natural and artificial selection as one unifying concept: only some individuals are favored to produce offspring.

So, the next time you see a pigeon or feed one in a park, just remember that Darwin used this humble bird to help support one of the greatest discoveries and theories of all time.

Use this T-Chart to **outline** the differences between natural selection and artificial selection.

Natural Selection	Artificial Selection

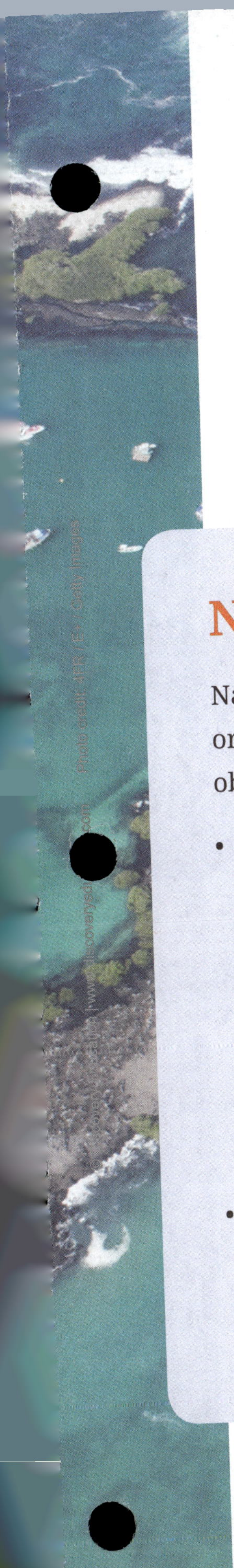

Activity 8

Analyze

Natural Selection at Work

Quick Code
ca8536s

Read the following text, and **watch** the video online. Then, with a partner, **discuss** how the finches that Darwin studied on his voyage could relate to each hypothesis.

EXPLORE

Natural Selection at Work

Natural selection is one of the mechanisms that enables species of organisms to change over time. Darwin based his hypothesis on four observations:

- All parents, regardless of species, can produce more offspring than are needed to replace them. For example, rats are found almost everywhere in the world. Rats are very adaptable and live close to humans, often surviving on their food scraps and other waste. Female rats commonly give birth to 10 young. They can produce a litter of baby rats every 3 weeks. Theoretically, a male and female rat, and their descendants, could produce 15,000 rats per year. That is a lot of rats! Why have rats not taken over the world?
- Conditions in the environment affect an individual's chances of survival. All organisms interact with their environment and rely on it to provide them with what they need to survive. They also must avoid being eaten. Many are not successful.

SEP Constructing Explanations and Designing Solutions

Natural Selection at Work *cont'd*

So, in any **population**, not all the individual organisms survive long enough to reproduce.

- Individual organisms are different. Look at a group of people. They all (apart from identical twins) look different. They exhibit what is called variation. Some organisms have features, or adaptations, that better suit them for survival.
- Some of these variations can be inherited. For example, moths inherit the color of their wings and people **inherit** the color of their eyes.

By combining these four observations, Darwin concluded that over many generations, better-adapted individuals, because they are more likely to survive, would increase in numbers within a population.

Darwin, the *Beagle*, and Finches

Describe how the evidence that Darwin collected on his voyage contributed to the theory of natural selection.

EXPLORE

Activity 9

Reason

Silly Selection

Quick Code
ca8538s

In this activity, you will mimic natural selection and create second and third generations of offspring with traits.

Predict

Which traits do you predict will be passed on to the next generation by natural selection?

SEP Developing and Using Models

CCC Stability and Change

What materials do you need? (per group)

- Seeds, black beans
- Seeds, kidney beans
- Seeds, lima beans
- Construction paper

HANDS-ON INVESTIGATION

Procedure

In this activity, you are the predator and the beans are your prey. Place 10 red beans, 10 black beans, and 10 white beans on the paper your teacher gave you. This paper represents your environment. These beans represent the first generation.

Pull as many beans as you can off the paper in 10 seconds. When the time is up, make note of how many of each color is left. Next, add two same-color offspring to the population for each pair of same-color beans. For example, if you are left with 2 red beans, 1 white bean, and 7 black beans, you will add 2 more red beans and 6 more black beans to the environment. This is your second generation. Make note of how many of each is left.

Repeat this for a third generation, and again make note of the results. Record how the phenotypic composition (that is, physical characteristics resulting from an organism's genetic content) of the generations changes over time.

EXPLORE

Reflect

How did you determine which organisms would be the parents?

How did the color of the environment (the paper) and the prey (the beans) impact the probability that a parent would be predated or would survive? Support your answer with data from your simulation.

How did the physical characteristics change over the generations?

How was the activity similar to and different from natural selection in nature?

EXPLORE

What Is the Connection between Genetics and Natural Selection?

Activity 10

Analyze

Quick Code
ca8539s

Genetic Variation in a Population

Read the following text. Highlight what an allele is, and then underline two examples of alleles.

Genetic Variation in a Population

A **population** is a group of organisms of the same **species** that can interbreed. For example, this could be the population of finches living on an island or mice living in a forest. Variation within a population of organisms is the raw material of natural selection. All organisms are constantly working to survive. For example, in a population of mice, each mouse is involved in searching for food, avoiding predators, finding a mate, caring for young, and much more. Any characteristic that helps with one or more of these jobs will aid the mouse's survival. For example, a mouse with a brown coat has a far greater probability of surviving on the forest floor than one with a white coat. If characteristics (or traits), such as a brown or white coat, are determined by the

Coat color varies within a mouse population.

mouse's genes, then in this situation, natural selection will select versions of those genes that produce a brown coat. Versions of genes are called alleles. Because natural selection works on all the organisms in a population, it can alter how common an **allele** is in that population. If natural selection favors brown mice but removes white mice from the forest population, the allele for a brown coat will become more frequent and the allele for a white coat will become less frequent. The allele for a white coat may, eventually, disappear from the population.

A population with a lot of **genetic variation**—for example, with many different alleles for different colors of mouse coat or finch beak size—will be more stable than a population with very little genetic variation. The reason for this is that most of the time, environments do change. A population with a lot of genetic variation will more likely have individuals in each generation that are adapted to the new conditions. Those adapted individuals will have more **offspring** than the average individual, and that **trait** will increase a little in the population. But when the environment changes back, other individuals will be better adapted. If there has been no or very little variation, then a sudden change in the environment would make every individual have fewer offspring and the entire population would decline. When genetic variation is low, a population may decline in the face of a changing environment; in extreme cases, it may disappear from an area, or the entire species may decline and become **extinct**.

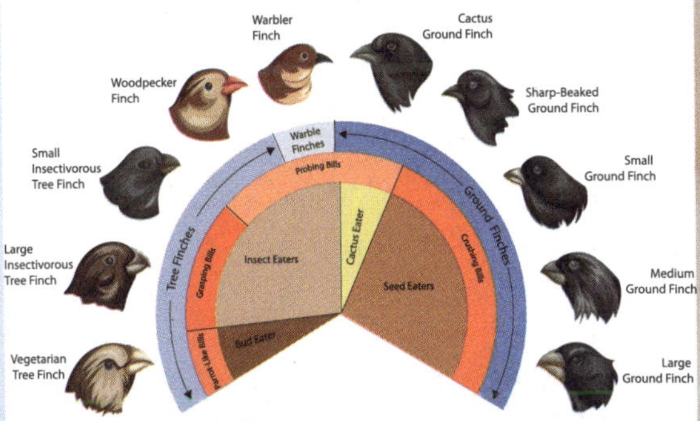

Beak size and shape varies within a finch population.

Activity 11

Reason

Birth Weight and Survival

Quick Code
ca8540s

In this investigation, you will analyze data on birth weights. Birth weight is the weight of a baby when it is born. Studies have shown that babies having certain birth weights are more likely to survive.

What materials do you need? (per group)

- Graph paper
- Writing utensils (pencils, regular and colored)

SEP Using Mathematics and Computational Thinking

CCC Patterns

CCC Stability and Change

Procedure

Table 1: Births in the Population

Birth Weight	Percent in Population
1.25–1.75	0.50
2.25–2.75	0.50
2.75–3.25	1.00
3.25–3.25	1.30
3.75–4.75	2.00
4.25–3.25	2.50
4.75–5.25	6.00
5.25–5.75	11.50
5.75–6.25	15.00
6.25–6.75	20.00
6.75–7.25	14.50
7.25–7.75	12.20
7.75–8.25	6.00
8.25–8.75	3.50
8.75–9.25	2.00
9.25–9.75	1.50

Table 2: Infant Mortality

Birth Weight	Percent of Mortality
2.0	70.00
2.5	50.00
3.5	30.00
4.0	25.00
4.5	6.0
5.0	6.8
5.5	3.0
6.0	2.0
6.5	3.3
7.0	2.8
7.5	2.0
8.0	4.0
8.5	3.5
9.0	3.8
9.5	4.0
10.0	7.0

EXPLORE

Review Tables 1 and 2 for data on birth weights and infant mortality. Then, create an investigative question that you want answered after the activity is completed. Determine which type of graph you will need to create to help you answer this question. Once your graph type is approved, create your graph on a separate sheet of graphing paper. Remember to label the axes and add a graph title. What conclusions can you draw from your graph?

Reflect

What is the shape of your Births in the Population graph? How would you describe the distribution of the birth weights based on your graph?

What conclusion can you make about how nature acts on low, intermediate, and high birth weight traits? Support your conclusion with data from your graph.

Why do you think there are fewer babies that survive in the extreme weight ranges?

Over the years, there has been an increase in the survival rate of low-weight and high-weight babies. What might be the cause of this increase?

EXPLORE

How Do New Traits Arise in a Population?

Activity 12

Observe

DNA

Quick Code
ca8541s

Watch this video, and **answer** the question.

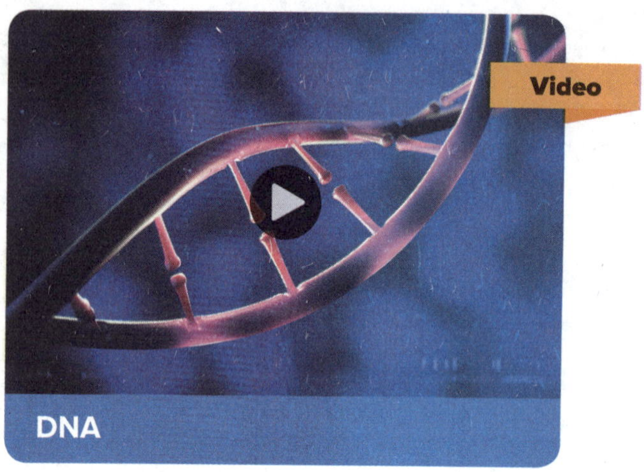

Why is DNA so important to living things?

____

____

____

____

____

____

Activity 13

Analyze

Mutations

Quick Code
ca8542s

As you read the following text, **highlight** the key ideas. When you are done, **select** one word that represents the big idea of this text. Then, you will **reread** this text and **underline** information about the relationships among DNA, genes, and proteins.

EXPLORE

Mutations

Genes are pieces of DNA. They contain hereditary information. This hereditary information allows parents to pass on physical characteristics to their offspring. These characteristics, called traits, are versions of genes that are located on chromosomes in the nuclei of eukaryotic cells. When an organism reproduces, the fertilized egg contains one set of chromosomes from the female parent and one set of chromosomes from the male parent. Each set of chromosomes has the same genes. But the version of a gene (an allele) may vary. For example, both parent mice will pass on genes for coat color. However, the actual coat color characteristics may vary depending on which allele is dominant.

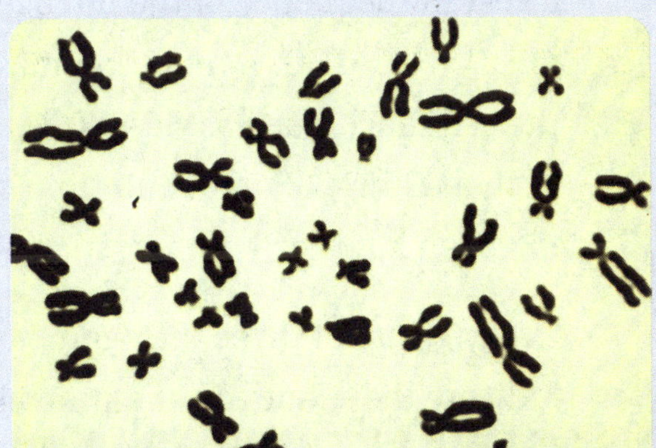
Chromosomes

CCC Structure and Function

CCC Stability and Change

Mutations *cont'd*

An allele of a gene contains the code that controls the production of one specific **protein**. This protein is made by the cell. Proteins are molecules made of long chains of amino acids. In a gene, groups of three nitrogen base pairs code for one of 20 specific amino acids. The order of these three base pair groups determines the order in which amino acids are assembled into proteins.

Proteins control all the cell's activities. They also determine the traits an organism has. For example, an allele for coat color codes for a protein that is responsible, in some way, for producing that coat color. If an allele is altered, then the protein may also change. This change may have no effect—be neutral—or it may cause a harmful or beneficial change in an organism. Changes in genes are called mutations.

A mutation is any change to the DNA of a gene. DNA mutations affect the sequence of the nitrogen bases. Consider this sequence of DNA: GCATCG. A mutation may cause a base to be added to the sequence, such as GCAATCG, or deleted from the sequence, such as GCATG. Some mutations involve the substitution of one nitrogen base for another, such as GAATCG.

Every group of three nitrogen bases is "read" by the cell to indicate the sequence of amino acids in a protein. The three nitrogen bases are called a codon. Changes to the order of the nitrogen bases of DNA can affect the order of the **amino acid** sequence in proteins by changing the codon. Changes can even affect the length of the protein if a mutated grouping is read as "stop adding amino acids." These changes can alter the shape and function of the protein.

Activity 14

Evaluate

Quick Code
ca8544s

Effect of Mutations on Proteins and Explaining Mutations

Codon	Amino Acid	Codon	Amino Acid	Codon	Amino Acid	Codon	Amino Acid
T T T	Phe	A T T	Ile	C T T	Leu	G T T	Val
T T C	Phe	A T C	Ile	C T C	Leu	G T C	Val
T T A	Leu	A T A	Ile	C T A	Leu	G T A	Val
T T G	Leu	A T G	Met	C T G	Leu	G T G	Val
T C T	Ser	A C T	Thr	C C T	Pro	G C T	Ala
T C C	Ser	A C C	Thr	C C C	Pro	G C C	Ala
T C A	Ser	A C A	Thr	C C A	Pro	G C A	Ala
T C G	Ser	A C G	Thr	C C G	Pro	G C G	Ala
T A T	Tyr	A A T	Asn	C A T	His	G A T	Asp
T A C	Tyr	A A C	Asn	C A C	His	G A C	Asp
T A A	stop	A A A	Lys	C A A	Gln	G A A	Glu
T A G	stop	A A G	Lys	C A G	Gln	G A G	Glu
T G T	Cys	A G T	Ser	C G T	Arg	G G T	Gly
T G C	Cys	A G C	Ser	C G C	Arg	G G C	Gly
T G A	stop	A G A	Arg	C G A	Arg	G G A	Gly
T G G	Trp	A G G	Arg	C G G	Arg	G G G	Gly

Effect of Mutations on Proteins

Use the genetic code to figure out the sequence of amino acids of a protein by completing the rest of the tables using the words in the word bank. If the codon codes for a "stop" instead of an amino acid, the protein ends at that point. Some words may be used more than once.

SEP Constructing Explanations and Designing Solutions

SEP Analyzing and Interpreting Data

EXPLORE

Word Bank

stop	Pro	Met	Tyr	Asp	Val

Gene

ATG	CCG	GAC	GTA	TAC

Mutated Gene

ATG	CCG	GAC	TGA	TAC

Explaining Mutations

How does a mutation cause a change in a protein?

What Is the Relationship between Natural Selection and Evolution?

Activity 15

Analyze

Natural Selection and Evolution

Quick Code
ca8545s

EXPLORE

As you read the following text, **look** for examples of cause-and-effect relationships. **Highlight** the causes in one color, and underline the effects.

Natural Selection and Evolution

A population has a common pool of alleles of genes—the gene pool. As new alleles (and sometimes new genes) enter a population by mutation or immigration, natural selection determines which persist and which disappear. For example, if the mutation is harmful, the mutated individual has a decreased chance of surviving and reproducing. If the mutation is beneficial, the individual survives to reproduce, and the mutated **allele** is passed to its **offspring**. Natural selection increases the frequency of these new useful alleles into the population and reduces the frequency of harmful alleles. The change in allele frequency within a gene pool is called evolution. Evolution therefore happens at the population level. But how does evolution produce new **species**?

CCC **Stability and Change**

CCC **Cause and Effect**

Natural Selection and Evolution *cont'd*

New species most commonly arise as a result of isolation. Imagine a lake with one population of guppy (a type of fish). The climate changes and the lake dries out to form two ponds. What was one population of guppies becomes two isolated populations. Their gene pools are separate. Mutations occur in each gene pool. Environmental conditions are slightly different between the ponds. Natural selection works on both populations, and the composition and frequency of alleles they contain begin to diverge. Over time, the two gene pools become very different. Guppies in the two populations begin to look and behave differently. The environment changes again and floods occur. The two populations mix. But now the members of each population are so different from one another they do not interbreed. They have become two distinct species.

The scenario above is one way that new species may form. It explains how natural selection can result in the formation of new species.

Activity 16

Evaluate

Quick Code
ca8546s

Speciation in Action

A species of butterfly moves through the process of speciation.
Label the events from 1 to 7, in the order in which they happened.

	A fire destroys the butterfly habitat in the valley.
	Butterflies from the western side and from the eastern side of the mountains return to the valley.
	Changes occur in the butterfly gene pools.
	Individuals in a species interbreed with one another in a valley between two mountains.
	Some butterflies escape to the western side of the mountains, while others escape to the eastern side.
	The butterfly habitat grows back in the valley.
	Two species of butterflies coexist in the valley but cannot interbreed.

CCC Structure and Function

CCC Stability and Change

CCC Cause and Effect

EXPLORE

Activity 17

Record Evidence

The Medium Ground Finch

Quick Code
ca8547s

As you worked through this lesson, you investigated and gathered evidence about evolution and natural selection. Now, take another look at the Medium Ground Finch image, which you first saw in Engage.

Let's Investigate the Medium Ground Finch

Read the Can You Explain? question from the beginning of this lesson.

Can You Explain?

What is natural selection, and how does it work?

Use your new understanding of the Medium Ground Finch to write a scientific explanation answering a question.

1. **Choose** a question. You may choose to answer the Can You Explain? question or one of the questions you wrote at the start of this lesson.
2. Use the following information to help you write your explanation.

SEP Constructing Explanations and Designing Solutions

Recall that a scientific explanation contains three elements: a scientific claim, evidence to support the claim, and reasoning that connects the evidence to the claim. As you plan your scientific explanation, be sure to include:

- reliable evidence from sources, including readings and/or videos
- data that provide evidence from investigations
- references to theories and/or laws that describe the natural world
- reasoning that shows how the evidence supports the claim

Record your notes in the spaces provided.

EXPLAIN

Using your notes, write your scientific explanation for the Can You Explain? question in the spaces provided here.

Scientific Explanation (Explanation = Claim + Evidence + Logical Reasoning)	
Directions: Create a scientific explanation for the question. This combines the claim that was made, the evidence that was collected, and a justification for why the evidence supports the claim (logical reasoning). The explanation should contain clear thoughts and accurate scientific vocabulary.	Key Components: • Uses precise and accurate language • Uses scientific vocabulary • Provides clear, logical thoughts • Uses evidence and reasoning to support the claim

EXPLAIN

STEM in Action

Activity 18

Analyze

Quick Code
ca8548s

What Is Natural History?

By the time Darwin sailed on the HMS *Beagle*, he had completed some training for a few different occupations. He had studied medicine, taxidermy, and theology—to name a few. Although the work he did may seem like that of a biologist, Darwin is more properly described as a naturalist. A naturalist is someone who practices natural history. Natural history is the observation and study of organisms in their natural environment.

Review the reading passage Voyage of Discovery to answer the questions that follow.

SEP Obtaining, Evaluating, and Communicating Information

Voyage of Discovery

Plymouth, England, December 27, 1831

A 22-year-old naturalist named Charles Darwin packs his bags and is ready to head off on a journey around the world. He'll be away for five years. The ship, called the HMS *Beagle*, is a research vessel, and Charles is joining the expedition as an unpaid observer. The living things and landscapes that Darwin would study, describe, and sketch in his notebooks would eventually change the way people look at nature and evolution.

HMS *Beagle* off Argentina, October 2, 1832

The journey wasn't always pleasant for Darwin because he suffered from bad seasickness. For days on end he stayed in his hammock, able to eat nothing but raisins. But one solution was to go ashore whenever he got a chance. Once the *Beagle* reached the coast of South America, there were many opportunities. On firm ground, Darwin observed everything around him. What particularly interested him was the fact that some species lived in one place but not another, even though the two places weren't that far apart.

In the morning we arrived at St. Fe. I was surprised to observe how great a change of climate a difference of only three degrees of latitude between this place and Buenos Ayres had caused ... In the course of an hour I remarked half-a-dozen birds, which I had never seen at Buenos Ayres. Considering that there is no natural boundary between the two places, and that the character of the country is nearly similar, the difference was much greater than I should have expected.

ELABORATE

Brazil, July 26, 1832

In Brazil, Darwin saw the biggest rodent he had ever seen in his life (in fact, it is the largest rodent species there is):

The largest gnawing animal in the world, the Hydrochaerus capybara (the water-hog), is here also common … In the daytime they either lie among the aquatic plants, or openly feed on the turf plain. When viewed at a distance, from their manner of walking and colour they resemble pigs … As I approached nearer and nearer, they frequently made their peculiar noise, which is a low abrupt grunt, not having much actual sound, but rather arising from the sudden expulsion of air: the only noise I know at all like it, is the first hoarse bark of a large dog. Having watched the four from almost within arm's length (and they me) for several minutes, they rushed into the water at full gallop with the greatest impetuosity and emitted at the same time their bark.

Galápagos Islands, September 15, 1833

As he filled his journals with sketches and descriptions of new plants and animals, Darwin continued to be amazed by the huge diversity of life. Different versions of the same animals seemed perfectly suited to their varied environments. Was there a connection?

It was most striking to be surrounded by new birds, new reptiles, new shells, new insects, new plants, and yet by innumerable trifling details of structure, and even by the tones of voice and plumage of the birds, to have the temperate plains of Patagonia, or rather the hot dry deserts of Northern Chile, vividly brought before my eyes.

Survival of the Fittest

By the time Darwin was 37 years old, he had published several works on his observations. His work revealed several major questions about living things: Why was there so much variation? Why did different species compete to survive? Why did some species become extinct? Did some species change? And the largest question he had of all: What was the origin of these species? For years, these questions haunted him, and the mystery of evolution remainedelusive. Then, in 1858, almost 30 years after his famous voyage, Darwin received a paper from a scientist named Alfred Russel Wallace. This paper contained many of the same theories that Darwin had considered. Just to see similar ideas in print helped Darwin put everything together. A year later, he published his most famous work, *On the Origin of Species*. Darwin put forth an amazing argument in this book: In nature, there exists a huge variety of species and environments.

In each environment there is a life-and-death struggle for food and for the means to reproduce. Without these, no species can last into the future. And only the most fit of each species survives: As many more individuals of each species are born than can possibly survive, and as consequently there is a frequently recurring struggle for existence, it follows that any being, if it vary in any manner profitable to itself, under the complex and sometimes varying conditions of life, will have a better chance of survival and thus be naturally selected. From the strong principle of inheritance, any selected variety will tend to propagate its new and modified form.

ELABORATE

Observations

Focus on Darwin's notes from Brazil on July 26, 1832. What details does Darwin include in his description of the capybara? What is the style and tone of his writing?

Notes

Imagine that Darwin visited your neighborhood to study the plants, animals, and environment. **Describe** a plant or animal species as Darwin might have seen it.

Notes to Theory

Darwin's work is based on his careful observation of plants, animals, and environments. What role do you think his note-taking took in his observations? How do you think his notes served him once he returned home? Are observation and note-taking still important tasks for scientists? Why or why not?

Activity 19

Concept Review

Review: Evolution and Natural Selection

Quick Code
ca8550s

Now that you have completed the objectives for this concept, **review** the core ideas you have learned. Record some of the core ideas below.

Core Ideas

EVALUATE

Talk with a Group

Now, think about the video Searching for a Cause that you saw in Get Started. Discuss how what you've learned about evolution and natural selection can help you understand the search for the cause of dinosaur extinction.

CONCEPT 3.3

Evidence for Evolution

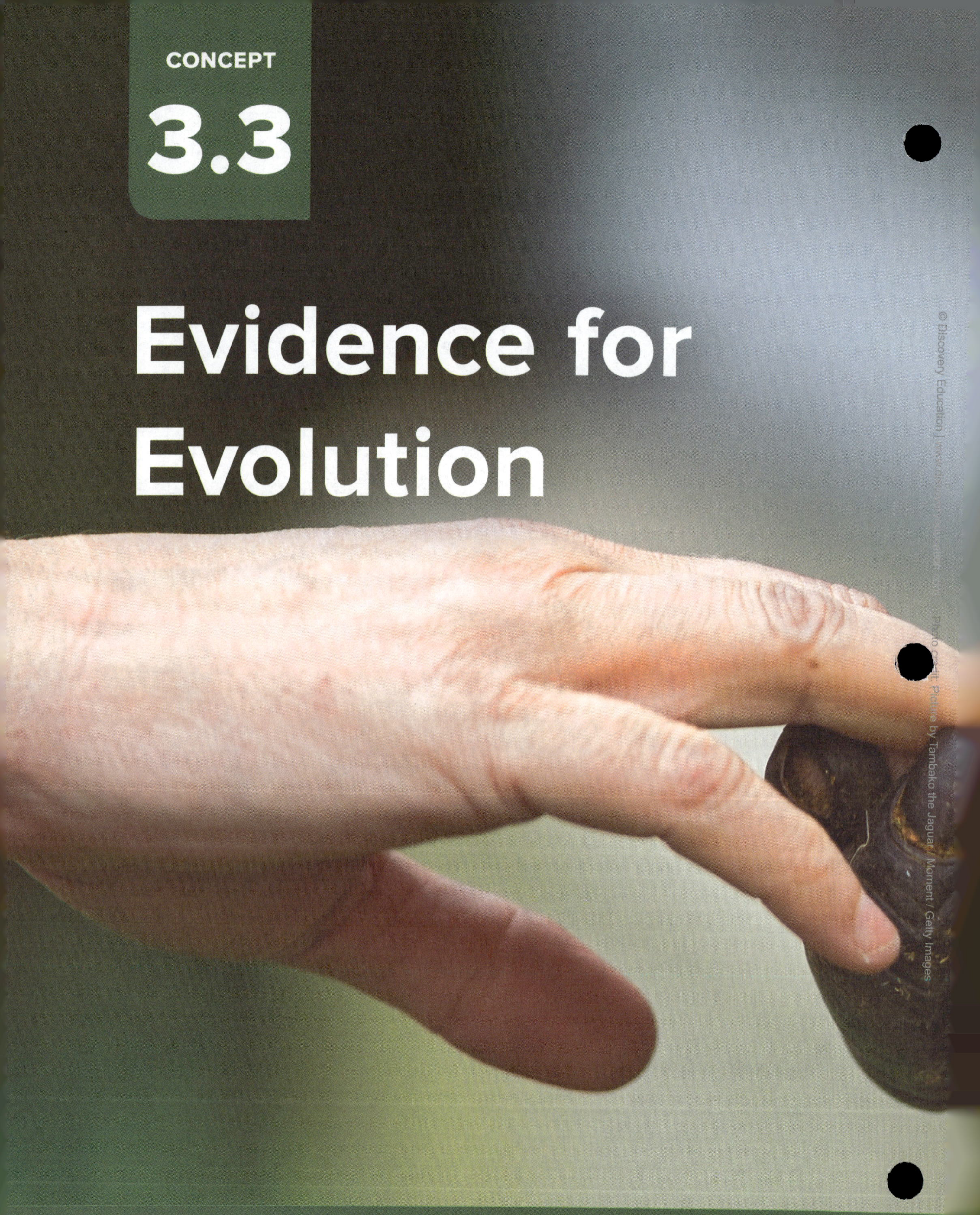

Student Objectives

By the end of this lesson:

- ☐ I can argue from evidence to explain why evolution is a scientific theory supported by evidence.
- ☐ I can analyze and interpret data from many sources on a variety of scales to construct explanations as to how the fossil record, homologous organs, embryonic development, and the presence of DNA provide evidence that evolution has occurred and is ongoing in the living world.
- ☐ I can develop models based on DNA and anatomical evidence that describe and predict the relatedness among organisms.
- ☐ I can construct explanations using evidence and reasoning for antibiotic resistance among bacteria and pesticide resistance among insects.

Key Vocabulary

antibiotic, backbone, DNA, disease, embryo, evidence, evolution, extinct, fossil, fossil record, germs, inherit, organism, plate tectonics theory, species, theory, theory of evolution, trait, vertebrae, virus

Quick Code
ca8552s

Activity 1

Can You Explain?

What kind of evidence do you think can be used to explain and support evolution in organisms, including humans?

Quick Code
ca8553s

Activity 2

Ask Questions

Flu Shots

Quick Code ca8554s

Have you had the flu? Most people have had flu, or influenza, at some time in their lives. **Look** at the image, and **think** about what you know about flu.

Let's Investigate Flu Shots

Write down a few questions you have about vaccinations.

ENGAGE

Activity 3

Analyze

Flu Evolution

Quick Code
ca8555s

Answer the following questions before reading the text. Then **read** the text, and **highlight** evidence that answers the questions.

What is the flu, and how does it affect your body?

What is a vaccine, and how does it work?

Why are new flu shots needed every year?

CCC **Structure and Function**

Flu Evolution

Have you had the flu? Most people have had flu, or influenza, at some time in their lives. Each year, students miss school with the fever and aches that come with this **disease.** In fact, influenza is a serious disease. The U.S. Centers for Disease Control and Prevention tracks flu outbreaks very carefully and produces a flu activity map that it updates weekly. Flu hospitalizes about a quarter of a million people every year. Each year, around 36,000 Americans die of the disease.

Flu is caused by a **virus.** You probably know about other viruses. The common cold is caused by a virus. However, flu produces much worse symptoms than colds do. Flu cases are more common in winter. Each fall, your doctor or local pharmacy advertises flu vaccines, often called flu shots. Flu vaccines are designed to reduce the chance of your catching the flu. You may have never had a flu vaccination. Chances are that you had one for another disease, perhaps measles, mumps, or polio. Usually vaccines last for a while—5 to 10 years or more—but for some reason, you need a flu shot every year. Why is that?

Vaccinations help the body resist diseases by telling its defense system how to make proteins called antibodies that fight the disease. When the virus attacks, the body is ready and the disease cannot take hold. This works great if the virus stays the same from one year to the next. However, the flu virus comes in many forms, or strains. Each year, scientists create a new vaccine based on what they expect the worst flu strains are likely to be for that year. The problem is that flu viruses keep changing. Viruses change through a process called natural selection. Every now and again, a new strain emerges or evolves into one that the vaccine does not affect. The flu virus is an example of **evolution** at work!

In this concept, you will learn about **evidence** for evolution, including how it can produce new forms of viruses and organisms.

Based on the class discussion, what is an additional question you could ask?

ENGAGE

Activity 4

Evaluate

Quick Code
ca8556s

What Do You Already Know About Evidence for Evolution?

Evolution Facts

Read the statements and **determine** if the statement is true or not true. **Write** "true" if it is a true statement and "not true" if it is not a true statement.

Sentence	True or Not True
Evolution most often takes a very long time.	
Humans continue to evolve, but other animals have stopped evolving.	
A theory is like an educated guess.	
Evolution is one of many theories that explain changes in human physiology, anatomy, and behavior over time.	

Humans and Other Apes

Some people think evolution claims that humans evolved from apes. This is inaccurate, however.

Primate Evolutionary Tree

What is true about the evolutionary history between humans and apes? **Use** the diagram to **help** you answer the question.

ENGAGE

Write two questions you have about human evolution based on the evolutionary tree.

Activity 5

Analyze

Both Fact and Theory

Quick Code
ca8557s

Read the following text about whether evolution is based on fact or is just a theory. As you read, **underline** information that supports the quote: "So, evolution is both fact and theory."

Both Fact and Theory

Evolution is a scientific **theory**. Like all scientific theories, it is backed up by numerous observations and measurements. So, evolution is both fact and theory. When scientists use the term *theory,* they mean something quite different from the common use of the word. In everyday use, the word *theory* means an idea used to explain something. Everyday theories may just be ideas backed up by little or no real evidence. Scientific theories are comprehensive explanations of a part of nature. In the case of evolution, the theory explains how life changes into new forms over time.

Evolutionary theory is one of many theories that attempt to explain how the universe works. You have come across many of them. Here are a few examples you may have heard of:

- Atomic theory explains that matter is made up of atoms.
- Gravitational theory explains interactions between bodies with mass—including planets and stars.

Photo credit: Juli[illegible]

- **Plate tectonics theory** explains how Earth is divided into giant tectonic plates.
- Germ theory explains how **germs** cause **disease**.

All these theories, including evolutionary theory, are backed up by huge amounts of evidence, and they are continually being tweaked and refined as new evidence arises.

Scientific theories can make predictions about the behavior of phenomena or even the existence of ones that have not been discovered. For example, gravitational theory can predict the behavior of the planets and of spacecraft, and evolutionary theory has predicted the location and existence of fossils and living organisms before they have been discovered.

Sometimes, evolutionary theory is depicted as being controversial. It is not. Because there is so much evidence supporting evolutionary theory, scientists no longer question it. Scientists know it has occurred and continues to occur. That does not mean they know everything about it. Scientists continue to make new discoveries that shed light on how evolution works.

You have already encountered some of the evidence for evolution. You have seen how the fossil record supports the view that life has changed over long periods of time, and observations that organisms change through the action of natural selection. In the rest of this concept, you will revisit some specific examples of this evidence, as well as other examples that have made evolutionary theory central to the science of biology.

Activity 6

Evaluate

Scientific Theory

Quick Code
ca8558s

Read the statements in the left column, and **determine** if the statement is an example of a theory or evidence that supports a theory.

Items to Classify	Examples of Theories	Evidence That Supports a Theory
explanation of what makes up matter		
fossil record of different animals		
description of how germs cause disease		
location of stars		
explanation of how Earth is structured		
behavior of planets		

How Does the Fossil Record Provide Evidence for Evolution?

Activity 7

Analyze

Quick Code
ca8559s

Evidence for Evolution from the Fossil Record

Read the text. As you read, **use** the graphic organizer to **note** the examples of cause and effect you find.

EXPLORE

Evidence for Evolution from the Fossil Record

The fossil record is our best resource for understanding past life, but it is incomplete. Some environments and some organisms did not form fossils well. During rapid periods of evolution and speciation, the likelihood of transitional species being preserved decreases. Only **species** that survived for a long time had good chances of being preserved or recorded in some way. It is likely that there are many species that have existed in Earth's past that humans will never know about. So fossils can only provide snapshots of evolution in action, much like a family photo album shows snapshots of children growing up. It doesn't show the complete picture, but it tells the story. However, scientists are continually finding new fossils that help fill in the gaps in the story. Most of the world's rocks have not yet been explored by paleontologists.

CCC **Cause and Effect**

Evidence for Evolution from the Fossil Record *cont'd*

What is most exciting is that as they do explore, they are finding more and more transitional fossils. And they are becoming better at predicting where to find them. Perhaps the best example is the location of fossils intermediate between reptiles and birds. The first of these was the famous Archaeopteryx, but recently, explorations in China have found hundreds of birdlike fossils. These and other recent discoveries have helped to unravel the evolutionary trees of many modern organisms. For example, the evolutionary tree for horses has been partially determined using fossils of many **extinct** horse-like species.

Another example is the evolution of whales. In the last few decades, scientists have uncovered part of the mystery of how whales, which are mammals, evolved as sea creatures.

Great efforts are being made to determine the place that human-like fossils have on the human evolutionary tree. Fossils, however, are not the only evidence being used to determine human ancestors. We can also look at our relationships to modern apes for additional clues.

Problem	Solution

EXPLORE

Activity 8

Observe

The Origin of Whales

Quick Code
ca8560s

Watch the video. During each pause, **work** with your group to **ask** a question or **make** a statement about what you just observed.

The Origin of Whales

SEP Constructing Explanations and Designing Solutions

CCC Stability and Change

CCC Structure and Function

Activity 9

Analyze

When Whales Walked

Quick Code
ca8561s

Read the passage. As you read, **highlight** information. On a separate sheet of paper, **draw** a flow diagram to summarize your findings.

EXPLORE

When Whales Walked

Whales, the biggest creatures on the planet, live in the sea, right? Sure, but it wasn't always this way. Fossils of whales show that these swimming mammals have been around for about 40 million years. They've been swimming in the oceans for a long time, but as far back as 60 million years ago, the ancestors of whales were actually walking on land.

You would think that whales evolved from fish because fish started out in the water. But the fact is that fish and whales aren't at all closely related. The reason evolutionists know this has to do with how fish and whales get their oxygen. Fish have gills that filter the oxygen out of the water. Whales have lungs and breathe the air. Another clue is how each reproduces. Fish lay eggs and leave them to fend for themselves, while whales give birth to live babies and nurse them until they're ready to be on their own.

SEP Constructing Explanations and Designing Solutions

CCC Stability and Change

CCC Structure and Function

But why did whales evolve to live in the water, if their ancestors lived on land like most other mammals? Let's consider what the species needed to do in order to survive. It was competing with other mammal species for limited supplies of food, and also it needed to be safe from predators. At one point, a key change took place, one that improved the whale ancestor's chances of survival to produce future generations. For some reason, this four-legged mammal that lived by the shore began spending more and more time in the water.

Picture an animal with strong front legs, long hind limbs, and a tail. Scientists call it *Ambulocetus natans*, and it lived about 50 million years ago. It was about the size of a sea lion, and it may have moved on land in a way similar to sea lions. It would have used its long tail and hind legs as paddles to move through the water. Maybe food was plentiful in the water and easier to catch. Perhaps the water provided this creature with a way to protect itself and escape predators. For whatever reason, over many generations, these seal-like walking whales took to the water. They became better swimmers and thrived.

Over millions and millions of years, back legs started to get smaller, tails morphed into flukes, and front legs became flippers. Ten million years after *Ambulocetus natans*, a creature known as *Rodhocetus kasrani* evolved. The water made these ocean swimmers feel almost weightless. As time went on, they got larger and larger, since the ocean water could help them support heavier weights. Hair became almost invisible, and a layer of insulating fat grew in and protected these warm-blooded creatures from cold water.

The whales as we know them began to emerge and split into two groups. One group kept their teeth—though they changed to adapt to the hunting of fish and other sea creatures. These whales became dolphins, porpoises, orcas, and sperm whales. The other group feasted on the smaller krill that was plentiful everywhere. Their teeth slowly developed into sheets of a horn-like mesh called baleen, which was perfect for sifting the tiny food. These whales became humpbacks, blue whales, and gray whales.

Out on a Limb

Evolutionists also draw conclusions about how species might be related by closely studying how different organisms are put together. They pay particular attention to structures that are similar. For example, what do you suppose you have in common with a whale and a bird? More than you think. We all share a common structure for the bones in our arms. All the forelimbs, or "arms," of these creatures have one "upper arm" bone, two "lower arm" bones, a series of small "wrist" bones, and several "hand" bones. A whale's fin has short, strong bones, but the pattern is the same as our own arms. This bone plan is exactly the same for all mammals, and it's similar for birds, amphibians, and reptiles. This bone plan is hard evidence that each creature is a variation on only one theme. Limbs have evolved in a variety of different ways to accommodate the changing environments that creatures live in. Humans have evolved to use their forearms not for walking or hanging but for grasping and carrying. No matter the variety that exists now, the fact that we all have the same number of bones in the same pattern proves that we all branched off the same tree. This means that we all share a similar ancestor.

EXPLORE

How Does Comparing the Structure and Development of Modern Organisms Provide Evidence for Evolution?

Activity 10

Observe

Homology

Quick Code
ca8562s

Watch the video. After watching the video, **write** the definition of the word *homology* and **provide** an example.

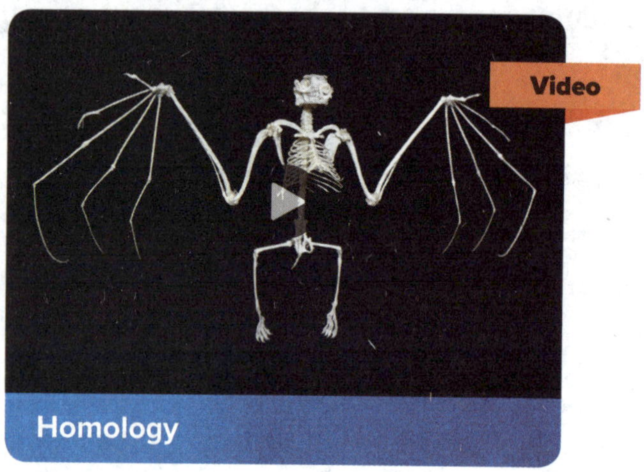

EXPLORE

Activity 11

Reason

Quick Code
ca8563s

Finding Homologous Structures

In this investigation, you will investigate evidence supporting the theory of evolution by comparing homologous structures in human and bird skeletons.

Predict

Do you predict that human and bird skeletons will be alike in any ways?

SEP Analyzing and Interpreting Data

CCC Structure and Function

What materials do you need? (per group)

- Skeletal drawings of human and bird
- Colored pencils

Procedure

Find the homologous structures in the human skeleton and bird skeleton. For each structure you find, use one color to shade in that structure on each skeleton. Use a different color for each structure you find.

When you are finished shading the structures, compare your drawings with those of a classmate. Did you both find the same structures? Add any that you missed.

EXPLORE

Reflect

What is a homologous structure?

Why would this type of evidence support the theory of evolution?

How would you design a different way to provide evidence that supports the theory of evolution?

EXPLORE

Activity 12

Observe

Quick Code ca8564s

Homologies in Development

Similarities in Embryological Development

Many organisms share similar traits during their early embryonic development. When examining the embryos of birds, fish, and mammals, you will find that there are many structures that appear in all of them. The simplest explanation for this phenomenon is that these groups all descended from the same ancestor species. While the adult forms of these creatures look quite different, the fact that they look very similar during embryonic development strongly suggests they have a common ancestor.

Watch the video. Then, **complete** the chart using words from the word bank.

CCC Structure and Function

Word Bank

amphibian	bird	fish	mammal	reptile

_______ _______ _______ _______ _______

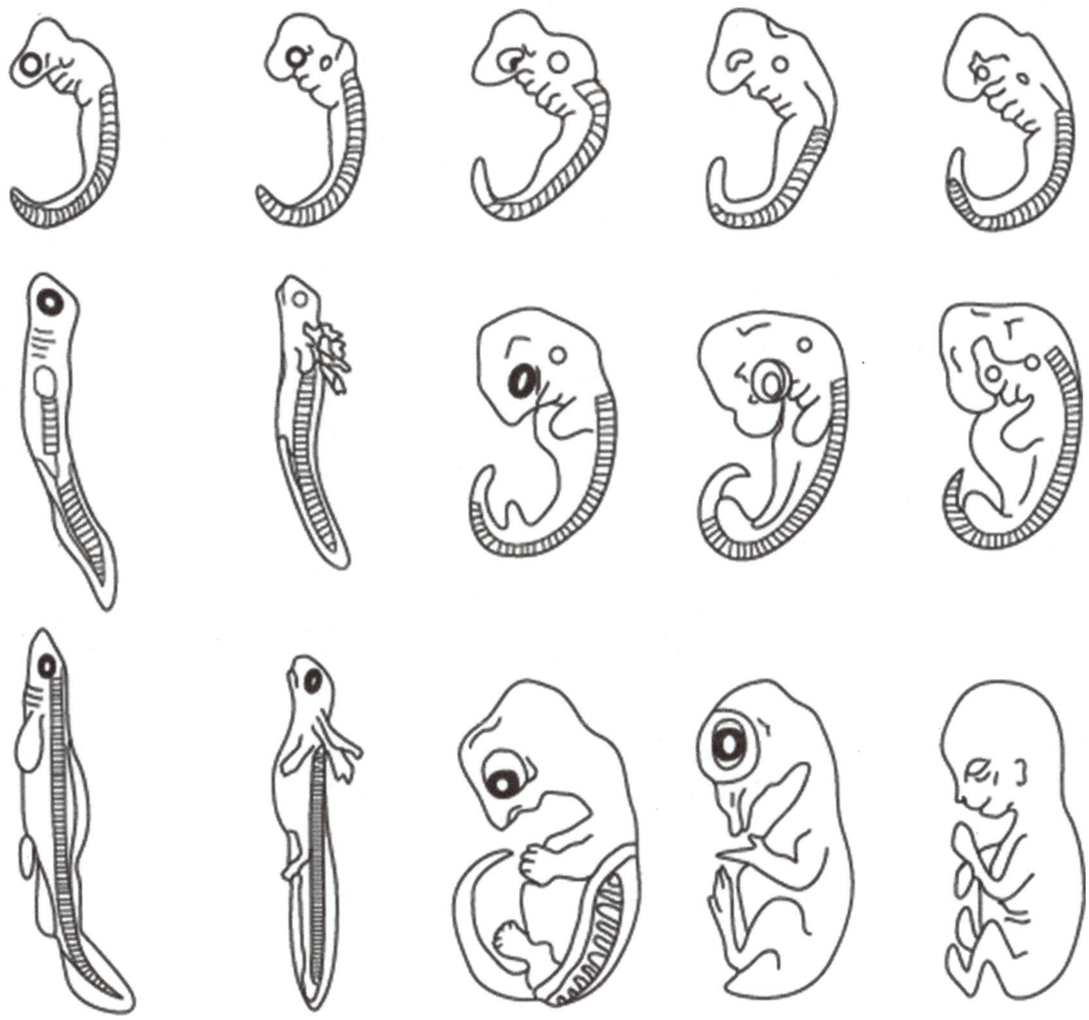

EXPLORE

Activity 13

Reason

There Have Been Some New Developments

Quick Code ca8573s

In this investigation, you will examine embryos to look for similar characteristics. Then, you will use these characteristics to see if these animals are related to one another.

What materials do you need? (per group)

- Photographs/diagrams of vertebrate embryos
- Markers
- Computer with Internet access
- Photographs/diagrams of insect larvae

Procedure

With your group, study the pictures of the animal embryos and human **embryo**. Take notes in your science notebook of the traits you see in each embryo. Then, list the traits that are common among all of the embryos and the traits that are different. Then, study the insect larvae photo. Compare it to the other embryos, and note any similarities and differences in your notebook.

When you are finished taking notes, discuss with your group what these pictures lead you to understand about evolution and common ancestry.

CCC **Structure and Function**

Reflect

What traits are common in all the species you observed?

What do these similarities tell you about the relationship among these organisms?

EXPLORE

How did the insect larvae photo differ from that of the other organisms?

Activity 14

Evaluate

Finding Homologies

Quick Code
ca8565s

Look at the arm structures of these organisms. Then, **answer** the questions.

Arm Bones

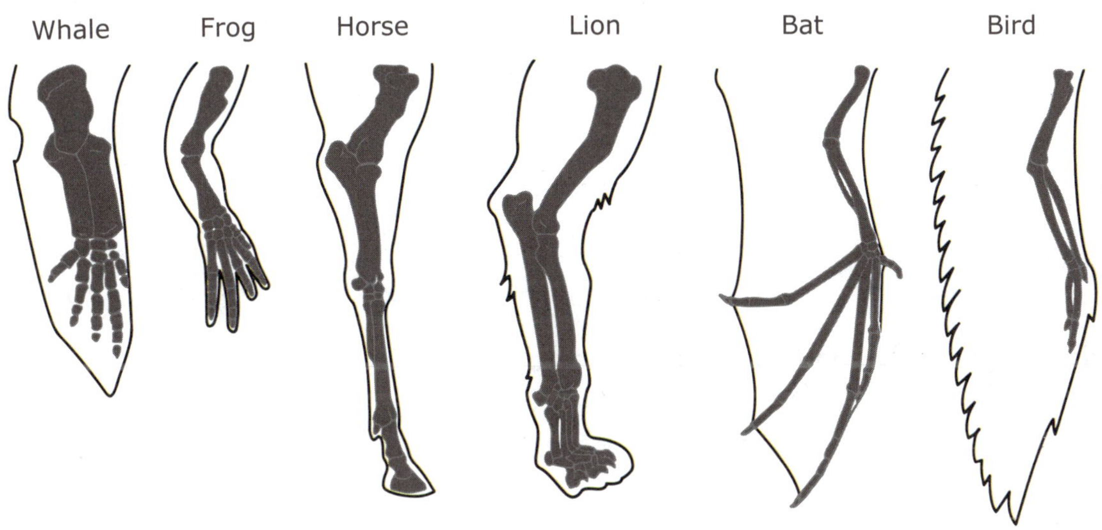

EXPLORE

SEP Constructing Explanations and Designing Solutions

CCC Structure and Function

CCC Cause and Effect

What is one similarity you see between at least two different organisms that you DO NOT see in the horse? **Describe** the structure and the organisms.

What do you think happened to the missing structure(s) in the horse over evolutionary time? Why?

Is There Any Genetic Evidence for a Tree of Life?

Activity 15
Observe

Quick Code
ca8566s

EXPLORE

Primate Cladogram

Study this cladogram. It shows the relationships among primates. Which primate is most closely related to humans? **Explain.**

Primate Cladogram

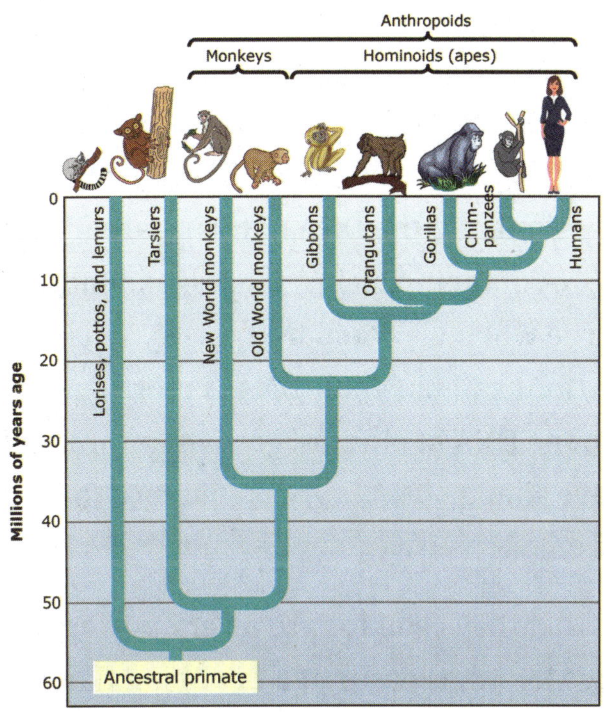

Activity 16

Analyze

DNA and Evolutionary Relationships

Quick Code
ca8567s

Read the following text. As you read, **highlight** the text that explains the role of genes as a tool in evolutionary biology.

DNA and Evolutionary Relationships

Because **DNA** is the blueprint for an **organism**, it should come as no surprise that genes can be used to trace ancestry. Shared DNA allows for certain traits to be inherited. We can map out the genes of any organism and compare DNA across species. This allows us to identify an organism's ancestors and relatives more easily than ever. An individual's DNA is very similar to the DNA of their parents. However, it is less similar to the DNA of their cousins, uncles, and more distant relatives. The same is true when we compare species. When we look at different lemur species, for example, their DNA is more similar to each other's DNA than it is to the DNA of chimps, gorillas, or orangutans. The fact that lemurs have similar DNA is great evidence that all lemurs shared a common ancestor.

Genes and the proteins that they code for are a very powerful tool in evolutionary biology. Mutations occur at a fairly standard rate. As mutations occur, they build up in a gene's DNA.

CCC **Structure and Function**

The number of differences in two similar genes in related organisms is, therefore, proportional to time because the organisms had a common ancestor. The genes of organisms carry an internal clock that records past evolutionary events and how closely organisms are related to one another. The use of changes in DNA to determine some evolutionary relationships has come up with some surprises. For example, by studying DNA, scientists discovered that flamingos, which look like storks and cranes, are, in fact, more closely related to pigeons than any other bird group.

Comparing all the genes—the genotypes—of different species has revealed that many organisms that are not closely related share very similar genes that do similar jobs. In fact, genes are often reshuffled or duplicated in organisms—a bit like you can reshuffle, add, or subtract different-shaped plastic construction bricks to make different models. All animals form a structure early in their development called a blastula. *Blastula* means "a hollow sphere of cells." These early stages of development of most animals are controlled by the same set of genes. The genes determine where and what type of structures, such as organs, eventually develop. These genes, also called Hox genes, are evidence that animals share a common evolutionary history. The reuse and shuffling of genes to form new organisms quickly generates a wide range of evolutionary possibilities.

Activity 17

Reason

Assemble a Cladogram

Quick Code
ca8568s

In this investigation, you will construct a cladogram and use it to show shared relationships among different organisms.

What materials do you need? (per group)

- Pencils
- Copy of the characteristics and taxa handout
- Copy of cladogram handout

SEP **Analyzing and Interpreting Data**

CCC **Structure and Function**

Procedure

Part 1: Creating a Venn Diagram of Shared Traits

1. Examine the table titled Characteristics and Taxa of Organisms. Make a list of the characteristics shared by most of the organisms, and then make a list of the characteristics shared by the fewest. Based on these lists, create a third list that orders the characteristics from most shared to least shared among organisms.

Part 1: Characteristics and Taxa of Organisms

Characteristic	Shark	Bullfrog	Kangaroo	Human
Vertebrae	X	X	X	X
Two Pairs of Limbs		X	X	X
Mammary Glands			X	X
Placenta				X

2. Construct a Venn Diagram using the third list. The characteristic shared by all taxa belongs on the outside of the diagram. The characteristic on the inside of the Venn Diagram should be the one shared the least among the taxa. Only the taxa that have a particular set of characteristics should be included in the Venn Diagram.

EXPLORE

Venn Diagram

Part 2: Creating a Cladogram

Construct the cladogram using the Venn Diagram you just completed.

Part 2: Cladogram Handout

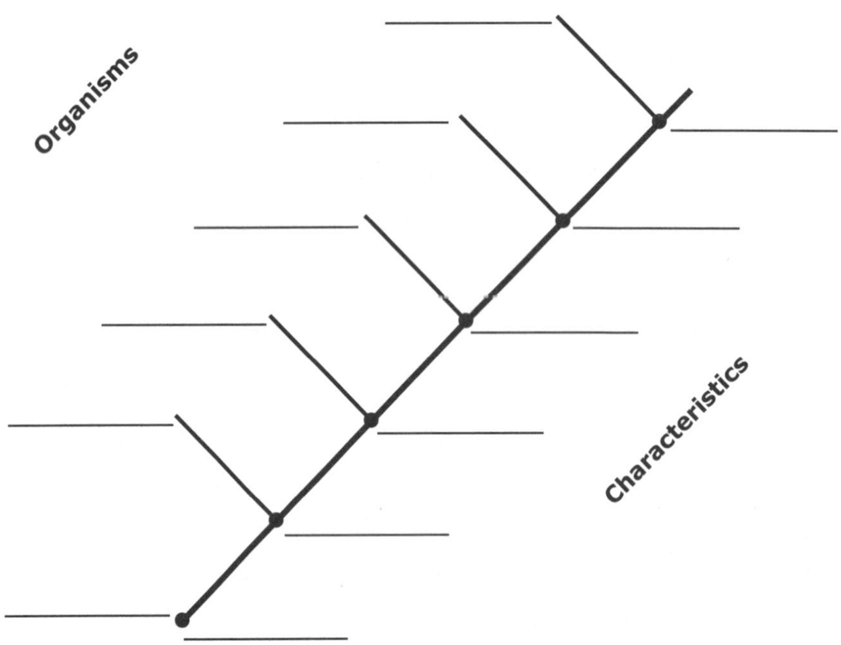

1. In the diagram, branches represent new species that broke off from their evolutionary ancestors. The nodes represent the characteristics that caused the branches to split. Label the nodes first. The characteristic shared by the most organisms belongs on the bottom node. The characteristic shared by the fewest organisms goes on the top node.

2. Label the branches of the cladogram with the organisms. The organism with the characteristics shared by the fewest organisms belongs on the top branch, and the organism with characteristics shared by the most organisms belongs on the bottom branch. The organism on the bottom branch represents the common ancestor of all the other organisms on the cladogram. All of the organisms on the cladogram share at least one common **trait** with this organism. The other organisms branched off from this common ancestor based on the development of new evolutionary traits.

EXPLORE

Reflect

Which of these characteristics is shared by the most of these organisms? Which of these characteristics is shared by the fewest of these organisms?

Which of these organisms has the fewest Xs in its column? Which of these organisms has the most Xs in its column?

What characteristic is shared between the common ancestor and all the organisms that evolved from the common ancestor?

EXPLORE

Activity 18

Analyze

Has Evolution Been Observed?

Quick Code
ca8569s

Read the section of text that is assigned to your group. Then, work with your group to **complete** the graphic organizer below with information from that section.

Claim	Evidence	Reasoning

CCC Cause and Effect

Has Evolution Been Observed?

Most of the evidence mentioned in this concept has been inferred from observations. Evolution, however, is ongoing and has been studied in real time. You have already encountered the study of the Galápagos medium ground finch evolving a longer beak in response to drought. Below are just a few more examples.

The Evolution of Pest Resistance

Pesticides are applied to crops to kill pests and weeds. Pesticides usually work well for a few years and then become increasingly less effective. This is because many pest populations have a few individuals that are resistant to the pesticide and survive an application. These resistant individuals breed, and over many generations, the genes for pesticide resistance spread throughout the population. Examples of such resistance include the resistance of malaria-carrying mosquitos to the insecticide DDT, the resistance of rats to some kinds of rat poison, and the resistance of weeds to the widely used weed-killer glyphosate.

The Evolution of New Viruses

Viruses attack the cells of the organism they are infecting. Once inside the cell, they replicate, killing the cell and releasing more viruses. This process is usually rapid, and many generations of virus can occur over a short period. Many viruses have very high mutation rates. Both these characteristics lead to rapid evolution. This can also lead to the viruses becoming resistant to drugs and vaccines.

The HIV virus (which causes AIDS) has become resistant to drugs through rapid evolution. If two different viruses infect the same cell, they can swap genetic material. These new mixed strains can cause new diseases—this is one way new flu strains arise.

The Evolution of Antibiotic Resistance

Antibiotics are drugs that kill disease-causing microbes called bacteria. When bacteria infect an organism, they reproduce rapidly—in a matter of hours. If a new mutation produces a bacterium that is resistant to the **antibiotic**, its offspring will become more numerous within the population. As a result, a new strain will arise that no longer responds to the antibiotic. The evolution of antibiotic resistance is a major problem; many antibiotics that were once very effective against certain diseases no longer work.

The Evolution of Faster Toads

Cane toads were introduced to Australia to control beetles that were eating the sugar cane crop. This proved to be a big mistake. The toads began invading the country and, because of the deadly chemicals they secrete, had no natural predators. Now, the toads have spread over almost half of the Australian continent. Scientists noticed that they were spreading faster as time went by. They discovered that those toads that reached new areas first had longer legs. Toads new to an area bred with each other—they had no choice. Their offspring had even longer legs. These toxic toads were evolving.

Evolution in Response to Pollution

Human activity has impacted nearly every environment on the planet. One of the most important impacts is pollution. Many organisms have evolved because of pollutants changing conditions in their habitat. Perhaps the best example of this is the peppered moth. The peppered moth is a species of light-colored moth that lives in Britain's woodlands and settles on lichen-covered trees. Its color camouflages it from bird predators.

During the 1800s, the air in some parts of Britain was badly polluted by smoke from coal fires. The smoke killed off the lichen and blackened the trees. The once-camouflaged moths became more visible to birds. However, like all populations, that of the peppered moth showed variation. One form of the moth that was rare in the population had darker wings. Birds found it more difficult to spot these dark varieties on the now-darker trees. As a result, over generations, the dark-colored moths began to outnumber the light. Within about 50 years, most of the moths in the populations living in industrial areas were black. About 50 years later, towns began to control pollution. The bark lost its soot, and the lichens began to return. Once again, the light form of the peppered moth was camouflaged, but now the dark form was not. The moth populations are now almost entirely of the light form. Natural selection worked in both directions, each time selecting the moth best suited to the color of the tree trunks.

EXPLORE

Activity 19

Reason

Superweeds

Quick Code
ca8570s

In this investigation, you will analyze a graph containing data from a study of ryegrass and how it responds to various herbicides. Ryegrass is a type of weed. You will analyze the data and form conclusions about how the ryegrass has evolved due to the herbicide.

Procedure

Study the graph and answer the questions.

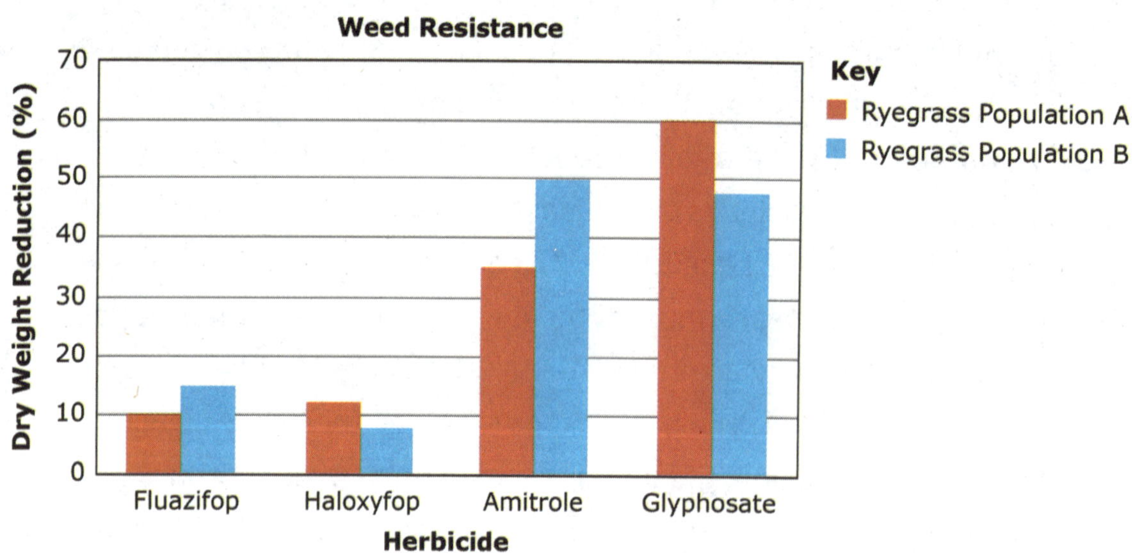

SEP **Analyzing and Interpreting Data**

What materials do you need? (per group)

- Graph of weed resistance to herbicides

How many different types of herbicides were evaluated for plant resistance in this study?

Which ryegrass population was the least resistant to herbicide treatment? Which was the most resistant?

Which herbicide was the most effective at treating the ryegrass populations? Which was the least effective?

EXPLORE

Reflect

What can be concluded from the results of this research?

Why was it important to evaluate the resistance of ryegrass to different herbicides rather than just one type?

How can you use the theory of evolution to explain why this ryegrass population is considered a "superweed"?

EXPLORE

Activity 20

Record Evidence

Flu Shots

Quick Code
ca8571s

As you worked through this lesson, you investigated and gathered evidence about evidence for evolution. Now, take another look at the image Flu Shots, which you first saw in Engage.

Let's Investigate Flu Shots

How has your understanding of the Flu Shots image changed?

Read the Can You Explain? question from the beginning of this lesson.

Can You Explain?

What kind of evidence do you think can be used to explain and support evolution in organisms, including humans?

Use your new understanding of the Flu Shots image to **write** a scientific explanation answering a question.

1. **Choose** a question. You may choose to answer the Can You Explain? question or one of the questions you wrote at the start of this lesson.
2. Use the following information to help you write your explanation.

You should use a scientific explanation to answer this question. Recall that a scientific explanation contains three elements: a scientific claim, evidence to support the claim, and reasoning that connects the evidence to the claim. As you plan your scientific explanation, be sure to include:

- reliable evidence from sources, including readings and/or videos
- data that provide evidence from investigations
- references to theories and/or laws that describe the natural world
- reasoning that shows how the evidence supports the claim

Record your notes in the spaces provided.

EXPLAIN

SEP **Constructing Explanations and Designing Solutions**

Using your notes, **write** your scientific explanation for the Can You Explain? question or your own question in the spaces provided here.

EXPLAIN

STEM in Action

Activity 21

Analyze

Career: Evolutionary Biologist

Quick Code
ca8572s

Read the text and **view** the videos. Then, complete the activity that follows.

Career: Evolutionary Biologist

You might think that there would be only one type of scientist for each type of living thing. After all, an entomologist studies insects, and a botanist studies plants, right? However, living things live in ecosystems and interact with one another in important ways. This means that the way each species evolves is not in isolation.

An evolutionary biologist is a scientist who studies the history of living things and how they evolved. The pathways of evolution for organisms in the same ecosystem are intertwined. One focus of an evolutionary biologist's work is to discover these connections. An evolutionary biologist is interested not only in how a species evolved but also in how that species may have affected the evolution of other species.

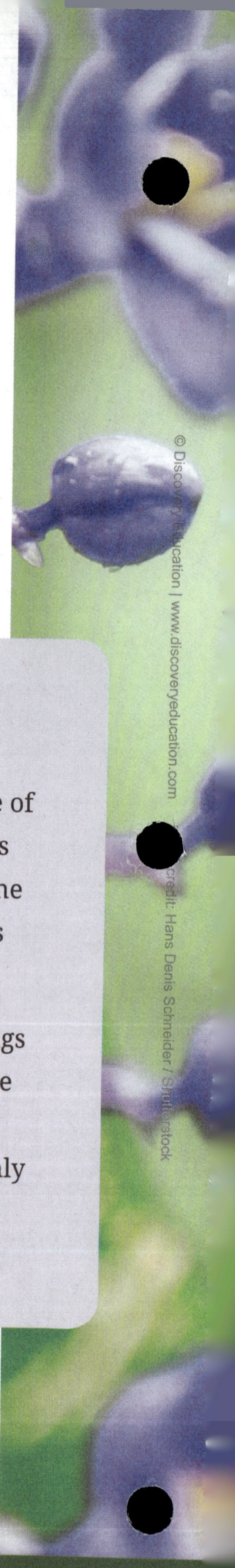

SEP Obtaining, Evaluating, and Communicating Information

The connections between the bee and the orchid are remarkable! How did the orchid develop such specific mechanisms and characteristics, ones that perfectly match the needs of the bee? Evolutionary biologists research and study this question and others like it.

Evolutionary biologists can track similarities and differences in the genes of related species to better understand how changes in these species developed over time.

Connecting behavioral changes to genetic changes is how evolutionary biologists show exactly how evolution occurred. Genes code for the processes and behaviors observed in a species. Discovering when these changes took place allows scientists to piece together just what happened.

Every species has an evolutionary history that is connected to the history of its environment. An evolutionary biologist uses observations and genetic data to identify what changes took place and how those changes led to the species of the present. With the right data, an evolutionary biologist also can predict what the species of the future might look like!

Video

Pollination from Animals and Genetic Evolution Improved Language Function

ELABORATE

Species Partnership

Research to obtain information about species partnership. **Use** an internet search for "coevolution" or "symbiotic relationships" to find examples of coevolved species, then **answer** the questions. **Use** the space provided to **show** your work.

What benefit does each species gain from its coevolution with the other?

How is the partnership evidence for evolution?

What do you think each species was like before the partnership, and what conditions do you think led to the partnership? **Formulate** a hypothesis based on evidence.

ELABORATE

Activity 22

Concept Review

Review: Evidence for Evolution

Quick Code
ca8574s

Now that you have completed the objectives for this concept, **review** the core ideas you have learned. Record some of the core ideas below.

Core Ideas

Talk with a Group

Now, think about the video Searching for a Cause that you saw in Get Started. Discuss how what you've learned about evidence for evolution can help you understand the search for the cause of dinosaur extinction.

EVALUATE

CONCEPT

3.4

Modifying Organisms

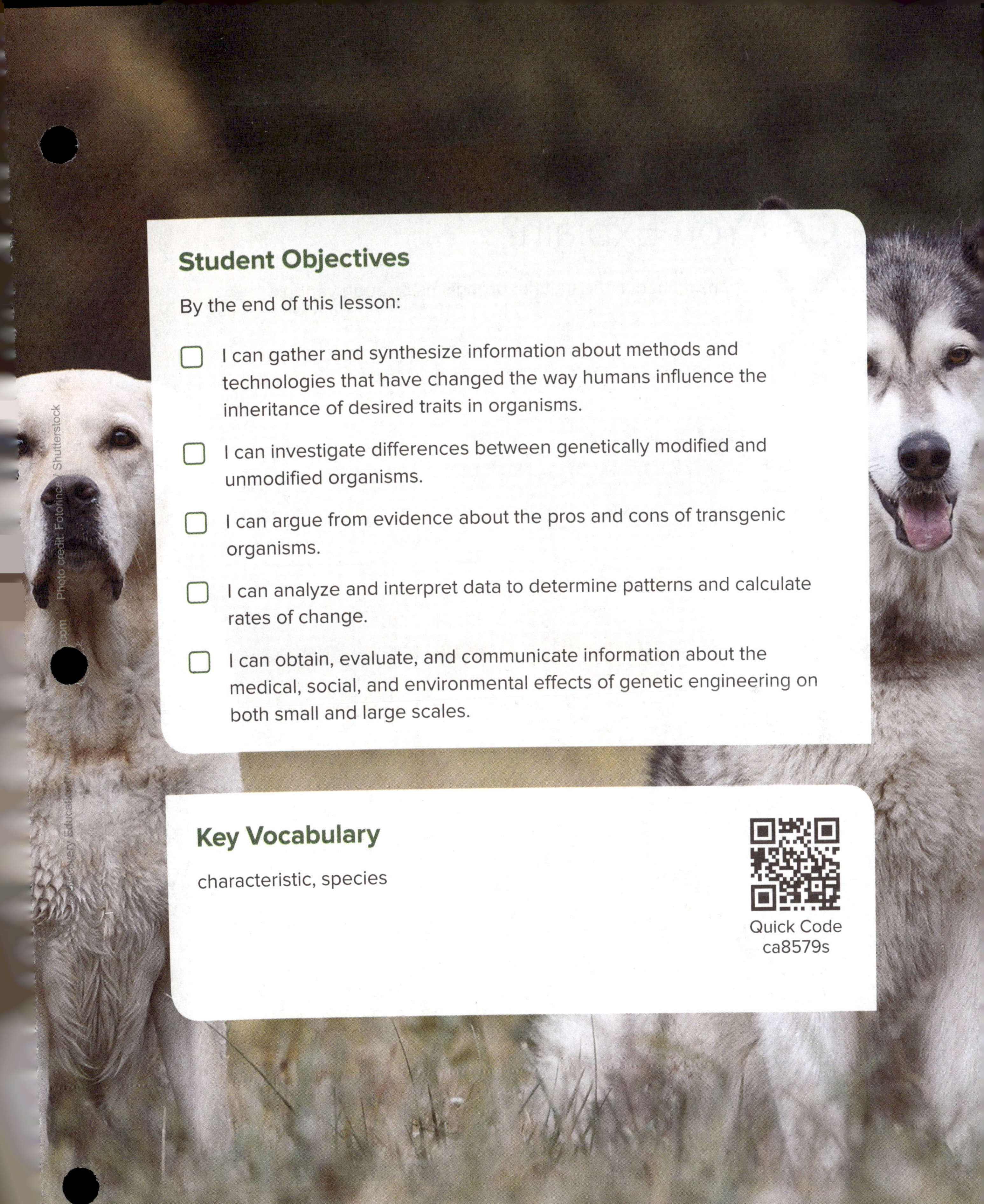

Photo credit: Fotorince Shutterstock

Student Objectives

By the end of this lesson:

- ☐ I can gather and synthesize information about methods and technologies that have changed the way humans influence the inheritance of desired traits in organisms.
- ☐ I can investigate differences between genetically modified and unmodified organisms.
- ☐ I can argue from evidence about the pros and cons of transgenic organisms.
- ☐ I can analyze and interpret data to determine patterns and calculate rates of change.
- ☐ I can obtain, evaluate, and communicate information about the medical, social, and environmental effects of genetic engineering on both small and large scales.

Key Vocabulary

characteristic, species

Quick Code
ca8579s

Activity 1

Can You Explain?

How do humans influence the traits of organisms through various techniques?

Quick Code
ca8580s

Activity 2

Ask Questions

Quick Code
ca8581s

Dogs

Some organisms have been domesticated. The process of domestication has been accompanied by artificial selection. Scientists can now modify organisms using a process called genetic engineering.

Let's Investigate Dogs

ENGAGE

What are some different dog breeds? **Identify** some structural and behavioral characteristics of these breeds.

Which of these traits are learned, and which are inherited?

Do you think these dog breeds would be able to survive in the wild?

What questions do you have about dogs and dog breeds?

Activity 3

Analyze

Thinking About Where Dog Breeds Come From

Quick Code
ca8582s

Using the image below, **describe** what you see. Then, **describe** what you think happened *before* the picture was taken, and **predict** what happened *after* the picture was taken. **Create** a Six-Word Story from the image. **Write** your story below. **Share** your responses with a partner, identifying any similarities and differences.

Dogs

CCC Structure and Function

ENGAGE

Read the text below. **Create** three different versions of your Six-Word Story that summarize the text. Working in groups, **share** your Six-Word Stories and **explain** your rationale. Each group should **select** or **create** what you believe is the “best” Six-Word Story and **share** it with the class.

Thinking About Where Dog Breeds Come From

Do you have a pet dog? Have you ever seen a dog show? Dogs come in many different shapes, sizes, and colors, and they have a wide range of personalities. Despite their vast differences, all our pet dogs, from the tiniest Chihuahua to the biggest Great Dane, are the same **species**, called *Canis familiaris*. The different types of dogs are called breeds. Some dogs are purebreds, coming from one specific breed, while others are mutts, coming from a mix of different breeds. Each breed has a distinct set of characteristics, and many exhibit **characteristic** behaviors. For example, terriers instinctively attack small rodents like rats and mice. Sheepdogs, like their name suggests, often work on farms herding sheep, and when they are kept as family pets, they will even try to “herd” small children. How were these dog breeds developed? Why do so many dogs seem to be made to do certain jobs for us?

It may be hard to believe when looking at a fluffy little Pomeranian or a poodle with a funny haircut, but our pet dogs are all descended from wolves. Animals that we keep as pets or as farm animals, descended from wild animal species, are called domesticated animals.

Thinking About Where Dog Breeds Come From *cont'd*

There are many different kinds of domesticated animals, and they serve a wide variety of purposes. They provide food, labor, and companionship for humans.

What exactly is a domesticated animal? What does it mean when we say an animal is bred for a certain purpose? How have humans "designed" these organisms? In this concept, you will learn about how humans influence the traits of animals and other organisms through breeding and other techniques. You will learn how, in some cases, humans actually can rewrite part of an organism's genetic code!

Activity 4

Evaluate

Quick Code
ca8583s

What Do You Already Know About Modifying Organisms?

Inherited Traits

A husband is talking to his pregnant wife, imagining what their child might be like. Which of the man's traits listed below could the baby potentially inherit from him?

- [ ] His dyed green hair
- [ ] His brown eyes
- [ ] The athletic build he grew up with
- [ ] The strong muscles he developed at the gym
- [ ] His freckles
- [ ] His suntan
- [ ] His tattoo
- [ ] His connected earlobes
- [ ] The hole from his ear piercing
- [ ] His index fingers that naturally bend away from each other

CCC **Cause and Effect**

Clones

Imagine a science fiction movie in which a scientist creates a new species of organism. The scientist generates a whole population of the organism entirely through cloning of the males and females, and then leaves them on an island. **Explain** why the population of clones would struggle to evolve if there were a major environmental change on the island.

ENGAGE

Activity 5

Analyze

Quick Code
ca8584s

Domestication of Organisms

Read the text describing the domestication of a dog. As you read, **highlight** text that you find interesting and **circle** text you find confusing. Then, **discuss** these parts with a partner: Did you agree? Disagree?

Domestication of Organisms

Have you eaten bread or popcorn today? Have you eaten a hamburger and French fries? If you have, you have eaten food from an animal or plant that humans have altered in some way. Almost all the food we eat—the main exception being many species of fish—comes from domesticated organisms.

Domesticated organisms are organisms that humans have altered by artificial selection, usually over many generations. This reproductive control by humans may have been accidental, intentional, or in most cases, a combination of both. Intentional selection of offspring of other species by humans is called artificial selection. In artificial selection, the selection by humans alters the probability of offspring surviving. Often, artificial selection selects for features that natural selection would

CCC Structure and Function

CCC Cause and Effect

eliminate. You may recall how Charles Darwin used artificial selection as a model for natural selection.

How does an organism become domesticated? Let's look at the dog as an example. Dogs have been domesticated longer than any other organism. They were originally a species of wolf and were probably domesticated around 30,000 years ago. We do not know the exact process of domestication; it probably happened more than once, and most likely it happened like this:

- A species of wolf lived in the same habitat as humans and spent time near the campsites of our ancient hunter-gatherer ancestors, eating scraps or bones.
- Some of these wolves were less aggressive than others and probably settled in the camp.
- Humans began deliberately selecting those that had behavioral and structural features that were useful.
- Humans continued to use artificial selection to breed dogs with different features so that they diverged into a series of dog breeds. The number of these dog breeds has rapidly increased over the past 250 years, giving us most of the breeds we recognize today.

It turns out that this pattern of domestication is common for many species—even plants.

EXPLORE

Chickens, bananas, corn, and turkeys have all been domesticated by humans. **Select** one to learn more about, either by reading or by viewing a video about the organism. As you research, take notes about the role that humans had in developing the organism's characteristics.

Domesticating the Chicken

Many people enjoy barbecue chicken or perhaps just having eggs for breakfast. Have you ever wondered when people started eating chickens and eggs? What were the first domesticated chickens like?

Scientists use several types of scientific evidence to answer these questions. Archaeological evidence shows us when and where people started eating chickens or using them for other purposes. Genetic analyses allow us to trace the chicken's ancestry back to its wild ancestors. Geneticists use mathematical models to estimate when genetic changes took place.

Scientists have learned that chickens primarily come from a colorful, wild bird called the red junglefowl, which still exists today in southeast Asia. This bird was hybridized with a closely related species, the gray junglefowl, which lives in India today. The wild birds were probably common around human settlements throughout Asia over the past millennium. However, scientific evidence suggests that people began domesticating chickens sometime between 5400 and 3600 BCE. The first chickens were probably domesticated in eastern China, but genetic studies show that they were probably domesticated independently in northeast India, southwest China, and southeast Asia. Over time, domesticated chickens were interbred, traded, and spread worldwide.

People bred chickens to enhance specific traits, depending on what the chickens were to be used for. Early uses included cockfighting (a sport in which male chickens fight each other to the death), food, and feathers for ceremonial purposes. Today, chickens are primarily used for meat and egg production, as well as showy plumage. These different uses have led to the development of hundreds of different breeds of chickens. For example, White Leghorns and Rhode Island Reds are popular breeds for use as laying hens. Plymouth Rock and Cornish are preferred meat chickens, and there are countless exotic and rare breed chickens for showing in competitions.

EXPLORE

History of the Banana

Bananas: The Same the World Over

Banana is the common name for the fruit and herbaceous plant that is part of the genus *Musa*. Bananas are one of the world's oldest and most popular fruits. They are very nutritious, generally inexpensive, and readily available. The banana plant is a large flowering plant that grows 6–7 meters tall. Each plant produces a bunch of bananas from a flowering stem.

Whether eating a ripe banana in the United States or in Europe, these store-bought bananas tend to taste the same. Part of the banana's popularity is due to its predictably delicious flavor. However, the uniformity that makes the banana so popular could also lead to its demise.

Banana History

As humans' hunter-gatherer ancestors roamed the jungle collecting food, they ignored the bananas they found. Wild bananas, which flower and reproduce sexually, produce hard seed cases with inedible seeds inside.

Occasionally, prehistoric humans found fruit on wild banana plants that did not contain seeds. These seedless bananas, when peeled, contained sweet, edible flesh. This is the edible banana that people know and enjoy today. Today's edible banana is a genetic mutation. The mutation produces tasty fruit but prevents proper seed development. The dark lines sometimes seen after biting into a banana are the stunted seeds.

Banana Sexual Reproduction

In nature, bananas reproduce through sexual reproduction. Sexual reproduction in flowering plants is similar to sexual reproduction in animals. Sperm cells are produced inside pollen grains. Pollen lands on the stigma, the female part of a flower. The pollen grows a long tube down to the plant's egg cell, and the pollen's nuclei move down the tube, fertilizing the egg cell. The developing seed contains the offspring, and the ovary begins to swell and develop into a fruit. Once again, the majority of bananas resulting from sexual reproduction are inedible.

EXPLORE

Cultivating Sterile Plants

Because humans' ancestors enjoyed the banana so much, they spent much time trying to domesticate the banana. They took cuttings from stems and replanted them. Cultivating plants by taking cuttings is a type of asexual reproduction. Asexual reproduction does not involve joining gametes.

Banana trees have a stem structure beneath the soil called a corm. Corms produce thin, fast-growing stems called suckers. When bananas are cultivated, the suckers are cut off, taking some of the thick corm as well. The sucker with the corm attached is planted in the soil, and a new banana tree grows. The new tree is identical to the parent tree.

Using tissue cultures for banana production is gaining in popularity. Tissue from banana plants is taken to labs to be cleaned and sterilized. The tissue begins cell division in nutrient solutions until new tissue forms. The new tissue is separated to make hundreds—even thousands—of new plants. The plants are clones of the parent. The offspring banana plants are disease free and grow prolifically.

The Importance of Sexual Reproduction

Sexual reproduction increases genetic diversity among individuals. If genetic diversity is low, there is low variation among individuals. Environmental changes in regions where banana plantations are located can be devastating due to the lack of variation among individual banana plants. If one plant in the plantation cannot adapt to changes, then none of the plants can, because they are all genetically identical.

Bananas Past and Present

Until the 1950s, the world's favorite banana was the Gros Michel variety. The Gros Michel was grown on plantations and cultivated by asexual reproduction using cuttings from parent plants. A root fungus called Panama disease quickly spread through plantations across the world, killing banana plants. Because the plants were genetically identical, they were not equipped to defend against Panama disease. The Gros Michel variety of banana is not extinct, but it is not part of today's global production. The Cavendish banana took over as the global banana in the 1960s. The Cavendish variety is naturally resistant to Panama disease.

Stalking the Cavendish

A fungal disease called Black Sigatoka is stalking the Cavendish banana. Plantations of cloned banana plants are incredibly vulnerable to disease, and fungus moves easily around the world. If a person stands on soil containing fungal spores, these spores can travel on that person's shoes and spread across the world within hours.

The natural world constantly battles disease. Diseases adapt and change. Organisms then adapt and improve resistance to overcome the diseases. Then the diseases adapt again, and so on. If a species like the banana has no genetic variety, its only hope for survival is intervention by science.

Are Bananas in Our Future?

It is too early to predict whether bananas will remain a mainstay of diets and economies worldwide. Recently, scientists announced they would be sequencing the genome of bananas. Scientists hope to unravel the entire genetic code in the DNA of banana cells. Once the genome is read and understood, genes within the banana's DNA that help disease resistance can be enhanced and altered. It may also be possible to modify the banana's genome by adding a gene for disease resistance from other plants, such as onions.

EXPLORE

Thanksgiving Turkey

Engineering and Growing Corn

Create presentation slides to demonstrate what you have learned. Then, **answer** the questions that follow.

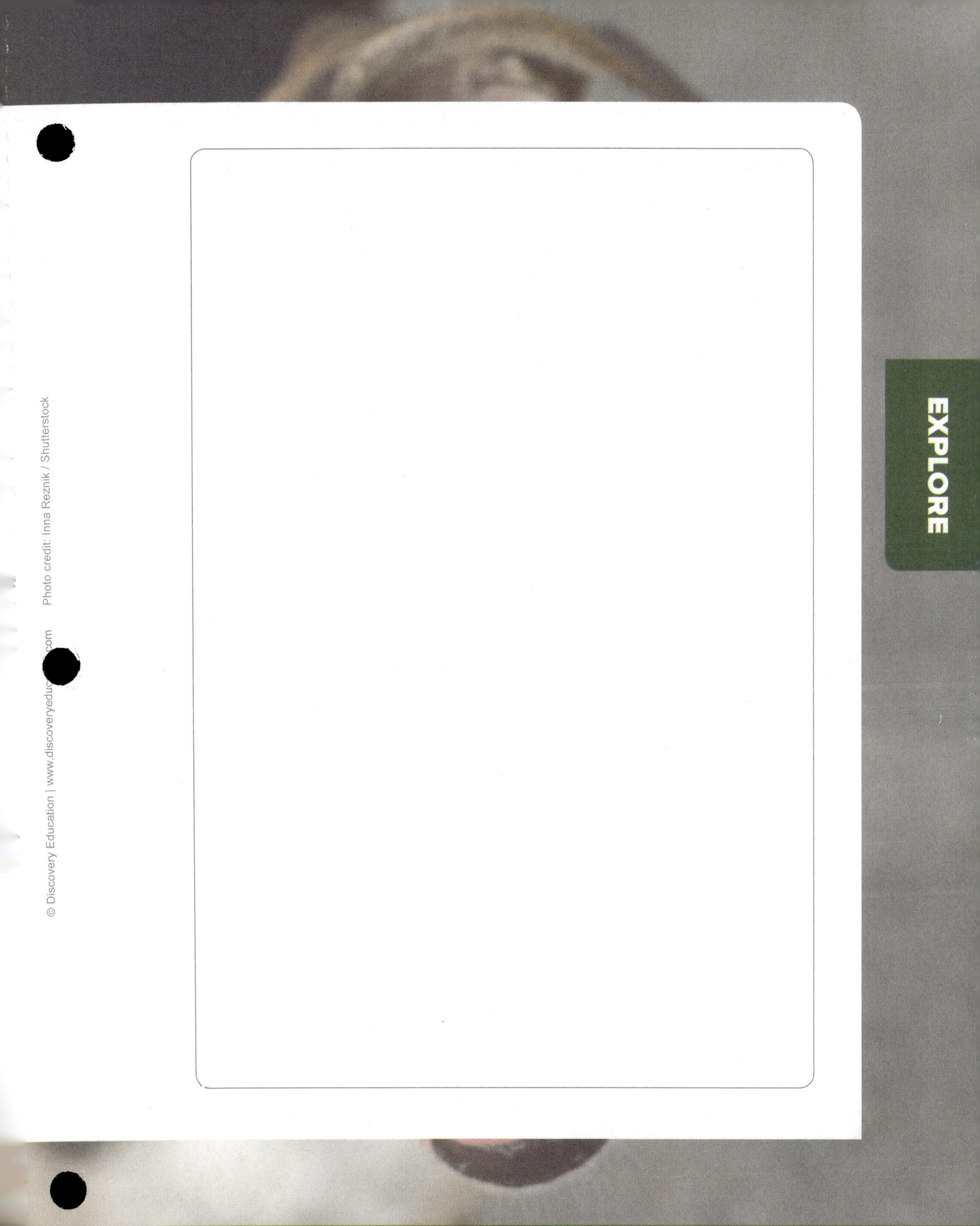

EXPLORE

How has domestication altered the distribution of these organisms?

How is domestication related to ancient civilizations?

What roles have domesticated animals played in human culture?

How have the animals benefited from domestication?

EXPLORE

What impacts have domesticated animals had on natural ecosystems and human populations in the past, and what impacts are they having today?

What Are Transgenic Organisms, and How Are They Used?

Activity 6

Observe

Quick Code
ca8593s

EXPLORE

Transgenic Animals

View the video Transgenic Animals. Then, **complete** the activities that follow the video.

What does it mean for an organism to be "transgenic"?

What is the role of transgenic organisms in the production of medicines and agriculture?

CCC **Structure and Function**

Work with a partner to **create** a T-Chart like this one to compare the pros and cons of transgenics. Make sure to **save** it, as you will use it in a later activity.

Pros	Cons

Activity 7

Evaluate

Quick Code
ca8585s

Changing Agricultural Practices, Rate of Increasing Corn Production, and Variable Yields

Changing Agricultural Practices

Based on the graph of average corn yields, when did corn-growing practices in the United States begin to change?

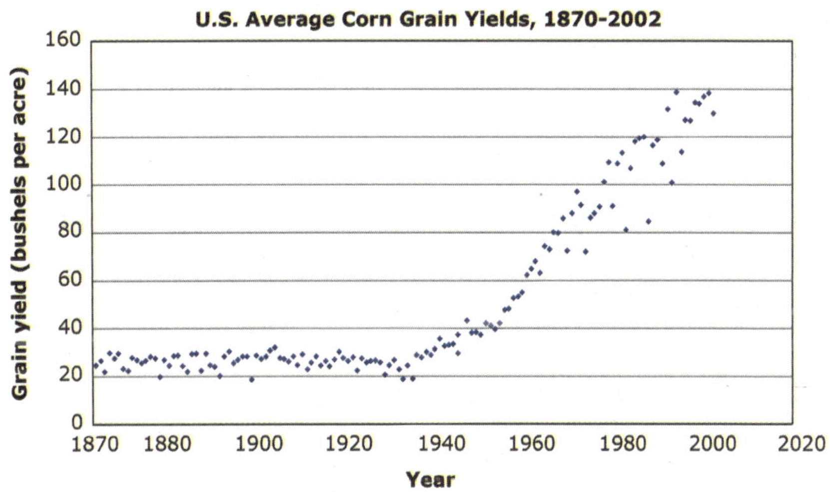

- [ ] A. Between 1900 and 1940
- [ ] B. Between 1940 and 1960
- [ ] C. Between 1960 and 1980
- [ ] D. After 2000

SEP Using Mathematics and Computational Thinking

SEP Engaging in Argument from Evidence

EXPLORE

Rate of Increasing Corn Production

Now, **work** in small groups of two or three to **brainstorm** how to calculate the rate of increase between 1950 and 2000. What was the approximate rate of increase between the two years?

- ☐ A. 2 bushels per acre per year
- ☐ B. 40 bushels per acre per year
- ☐ C. 90 bushels per acre per year
- ☐ D. 140 bushels per acre per year

Variable Yields

The graph shows that corn production has varied since 1970. (The data points are more scattered.) Which explanations could account for this? Select all that apply.

- ☐ A. Some farmers do not have advanced GIS technology for planting and managing their crops.
- ☐ B. Corn is grown in more places with different soils and climate than before.
- ☐ C. Farmers are no longer crossbreeding corn.
- ☐ D. Corn is now being cross-pollinated with other types of plants.

As a class, **decide** on the production values you will use for the two years, 1950 and 2000. **Work** with a partner to **make** the calculation. **Share** the steps you took with the class.

What Social and Environmental Impacts Result from the Use of Genetic Engineering?

Activity 8

Evaluate

Quick Code
ca8586s

EXPLORE

Engineering Organisms

Work in groups of two or three. **Read** the statements in each of the following cards. **Label** each card either *artificial selection* or *genetic engineering*. Then, **write** two of your own cards, providing one example of genetic engineering and one example of artificial selection. **Label** your new cards correctly.

A farmer picks a corn plant that has the sweetest corn and crosses it with a corn plant that makes many ears of corn to get a plant that makes ears of sweet corn.	A scientist alters the DNA of an apple to prevent the apple from turning brown after it is cut.
______________________	______________________
An Atlantic salmon has a gene from an eel that allows it to grow year-round.	A goat contains DNA from a spider that allows silk to be made from its milk.
______________________	______________________

A farmer mates the largest male and female chickens in order to produce larger chicken offspring.

A gardener crosses a rose plant with very few thorns with another rose plant with few thorns.

A rose plant has DNA from a pansy plant that gives the rose a blue flower.

A rice plant contains DNA from carrots that give the rice a golden color and make it produce beta-carotene, which must be converted to vitamin A before it can be used by the body.

Activity 9

Analyze

Social and Environmental Impacts of Genetic Engineering

Quick Code
ca8587s

Read the text. **Highlight** additional information that you can add to your T-Chart that you created in your notebook to compare the pros and cons of transgenics.

EXPLORE

Social and Environmental Impacts of Genetic Engineering

All new technologies come with benefits and risks. Some of the benefits are already being realized; others are likely to come sometime in the future. They include larger harvests of crops, longer shelf-life of products (one of the first marketed genetically modified organisms, or GMOs, was a tomato with a long shelf life), the improved nutritional value of certain crops, and the low-cost production of medicines. Genome mapping and gene therapy—the replacement of defective genes with healthy ones—have enormous potential for understanding, treating, and preventing disease. Scientists have successfully created human embryos with selected genes and implanted them into a woman's uterus. The genes were selected to avoid genetic disorders. However, many people are concerned about the implications. Could genetic selection lead to a future in which people select exactly the traits they want to see in their children? Is selecting genes for blue eyes, athleticism, or red hair the same as selecting genes that will prevent disease? Is it right to manipulate human genetics at all? These questions are complicated. Scientists must always consider the ethical implications of their research.

One potential use for genetic engineering is to help save endangered species. Scientists can harvest and replicate DNA from tissue samples of endangered species. This DNA is frozen and stored for future use in cloning. Cloning is one possible method of increasing the population of endangered species. It could also be used to reinstate extinct species. However, this, too, raises some ethical questions. How do we decide which species to save? Who should be responsible for saving them?

So, what are the risks? Currently, there is no evidence that food substances such as starch, sugar, or oils derived from genetically modified crops are harmful. We do know, however, that genetically engineered crops have the potential to cause health problems. For example, in lab tests, soybeans with a nut gene—added to improve their quality as animal feed—produced allergic responses in people with nut allergies. Genetically modified crops, as another example, can cause environmental problems. Plants genetically engineered to kill pest insects may also prove toxic to non-target and valuable species such as butterflies and bees. Additionally, genes integrated into the genomes of crop plants can inadvertently become incorporated into closely related non-crop organisms through pollination or by disease-causing organisms. In this way, “foreign genes” may enter and cause harmful changes in natural ecosystems.

The biggest problem with genetically modified crops, so far, has been indirect, through the production of pesticide resistance—particularly resistance to weed killers. Agricultural companies have developed crops that are resistant to weed killers—the idea being that these chemicals could be applied and kill only weeds and not crops. This has indirectly led to the natural selection of resistant superweeds.

EXPLORE

Activity 10

Observe

Genetic Selection

Quick Code
ca8588s

View the video segment Genetic Selection. **Work** in pairs to **debate** genetic engineering, including human genetic engineering, with one group in favor and the other group opposed. **Use** the debate strategy Talking Stick to **designate** the speaker and a set amount of time during the debate.

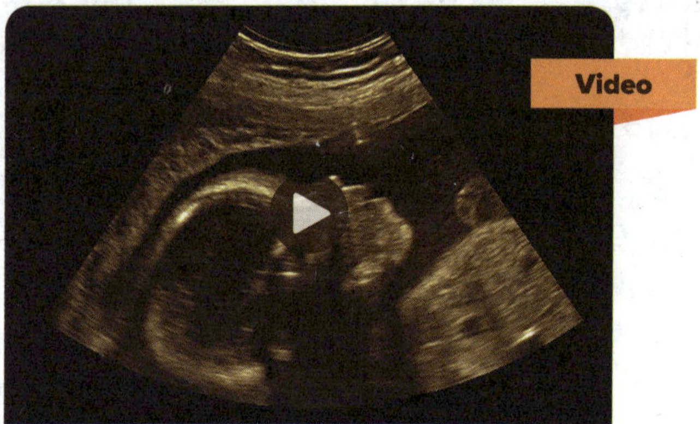

Genetic Selection

SEP Engaging in Argument from Evidence

CCC Cause and Effect

Activity 11

Investigate

Hands-On Investigation: GMO Test

Quick Code
ca8589s

In this investigation, you will perform tests to determine whether individual corn plants have been genetically modified to resist a certain herbicide.

Predict

What do you predict will be the difference between genetically modified and unmodified corn plants?

SEP Constructing Explanations and Designing Solutions

CCC Structure and Function

CCC Cause and Effect

EXPLORE

What materials do you need? (per group)

- Organic corn seeds (divided evenly between two envelopes), 10
- Corn seeds genetically modified to resist the herbicide glyphosate, 10
- CP4 EPSPS test strips, 4
- Soil, potting
- Large flower pots, 4
- Labels
- Markers
- Water dispensing bottle, 30 mL
- Water
- Safety goggles (per student)
- Gloves (per student)
- Lab apron (per student)

Procedure

Part 1: Background

Read the following text before beginning your investigation. View the videos for more background on this subject.

Direct gene modification techniques are significantly more modern than other artificial selection techniques (like selective breeding), and consequently some scientists define genetic engineering to include only gene modification techniques. Geneticists are actively working on advanced artificial selection techniques, and some scientists do not consider these techniques to be part of genetic engineering.

As a result, there can be some debate as to which foods are and are not considered genetically modified. Nearly all of our modern crops and livestock are the result of centuries of artificial selection and domestication, but many of these are not considered genetically modified unless they have been subjected to direct gene modification techniques—for example, transgenesis.

Test strips test for the presence of an enzyme called CP4-EPSPS. This is a special enzyme taken from herbicide-resistant bacteria and added through transgenesis to corn, soybeans, tobacco, and other crops. A plant is only considered organic if it has had no gene modification.

Part 2: Guided Inquiry

Use the lab materials and your background knowledge from Part 1 to proceed with your investigation. Respond to the following questions to guide you.

How can you use the lab materials to create an investigation to determine which of these seeds are from GMO crops and which are from non-GMO crops?

EXPLORE

How can we use these materials to create an investigation to determine which of these seeds are from herbicide-resistant crops and which are from non-herbicide-resistant crops?

How can you avoid bias in your tests?

Part 3: Design and Perform an Investigation

Using your group's unknown numbered seeds—either organic seeds or GMO seeds—develop your lab procedure to determine whether your group's seeds are genetically modified to be herbicide-resistant.

Recall in Part 1: Background that the special enzyme CP4-EPSPS is taken from herbicide-resistant bacteria and added through transgenesis to corn, soybeans, tobacco, and other crops. Form your group hypothesis. Prepare and plant the seeds in the pots, making sure to label pots with group members' names. Place near the windows, and allow time to grow.

Part 4: Perform and Complete an Investigation

After your group's plants have had adequate time to grow (at least two weeks), complete your investigation according to the procedure your group has designed.

Design your own method to analyze your data to answer questions and compare the results to their hypothesis.

EXPLORE

Examine the plants and discuss whether the group thinks the plants are GMOs or organic. Explain your reasoning. Record the group's responses in your notebook.

Once all responses are recorded, discuss the hypotheses as a class and use the test strips to discover which plants were actually organic and which were GMOs.

Reflect

Compare your results to your hypothesis. Was your hypothesis correct?

Has this experiment changed or reinforced your personal opinion about GMOs? Why or why not?

Does a negative test result from the CP4-EPSP-synthase test strip prove that a plant is not a GMO? Why or why not?

Why might people be concerned that these types of GMOs might lead to an increase in herbicide use?

EXPLORE

Activity 12

Record Evidence

Dogs

Quick Code
ca8590s

As you worked through this lesson, you investigated and gathered evidence about modifying organisms. Now, take another look at the image Dogs, which you first saw in Engage.

Let's Investigate Dogs

How has your understanding of the Dogs image changed?

Read the Can You Explain? question from the beginning of this lesson.

Can You Explain?

How do humans influence the traits of organisms through various techniques?

Use your new understanding of Dogs to **write** a scientific explanation answering a question.

1. **Choose** a question. You may choose to answer the Can You Explain? question or one of the questions you wrote at the start of this lesson.
2. Use the following information to help you write your explanation.

You should use a scientific explanation to answer this question. Recall that a scientific explanation contains three elements: a scientific claim, evidence to support the claim, and reasoning that connects the evidence to the claim. As you plan your scientific explanation, be sure to include:

- reliable evidence from sources, including readings and/or videos
- data that provide evidence from investigations
- references to theories and/or laws that describe the natural world
- reasoning that shows how the evidence supports the claim

Record your notes in the spaces provided.

EXPLAIN

SEP **Constructing Explanations and Designing Solutions**

Using your notes, write your scientific explanation for the Can You Explain? question in the spaces provided here.

EXPLAIN

Share your scientific explanation with a partner. **Read** your partner's scientific explanation, then **complete** the table below.

Record the writer's claim.
List three pieces of evidence the writer used to support the claim. Put a star next to evidence different from your scientific explanation. 1. 2. 3.
Interpretation of the Facts – What facts were interpreted differently? Do you agree with the interpretation? Why or why not?

EXPLAIN

STEM in Action

Activity 13

Analyze

Careers in Genetic Engineering

Quick Code
ca8591s

Read the text and **complete** the activity that follows.

Careers in Genetic Engineering

You may know that fossil fuels are a nonrenewable resource. Eventually, they will become scarce. Many alternative fuels are being developed. However, none are yet as powerful as fossil fuels for meeting our ever-increasing energy demand. This is a major reason that we are still very resistant to giving up fossil fuels, even though we know we will eventually have no choice.

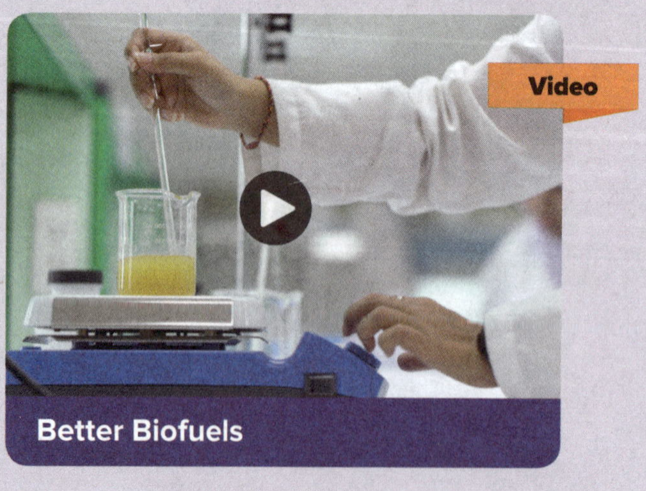

Better Biofuels

SEP Obtaining, Evaluating, and Communicating Information

Synthetic biologists are trying to solve this problem through genetic engineering. Microorganisms, or microbes, already have an extremely wide variety of metabolisms. They also have remarkable abilities to change the chemistry of the environment around them. Synthetic biologists are working to harness these abilities. These scientists are working to genetically engineer microbes to produce chemicals that are very similar to the fuels we already use. If synthetic biologists could achieve this goal on a large scale, we could replace the gasoline and ethanol we use now with a chemically similar fuel. This fuel, however, would be from a renewable source.

New alternative energy sources are not the only area of focus for genetic engineers. Geneticists are working on developing new crops. Some geneticists focus on crops to improve our food supply, but others have different goals. Some geneticists are engineering something as simple and common as corn to turn it into life-saving medicine.

Spider Silk

ELABORATE

Careers in Genetic Engineering *cont'd*

As with all genetic engineering, there are some risks involved. Scientists must prevent experimental plant strains from crossbreeding with other crops that may end up in the food supply, which could have serious, but unknown, consequences. To prevent this from happening, scientists have to make some sacrifices when it comes to creature comforts on the job. Working in a deep underground lab may not be as nice as working in a corner office with a view of a city skyline. However, it is worth it for the important work they are doing and to protect the safety of the public.

Transferring genes from one microbe to another or crossbreeding different varieties of corn is incredible. It also sounds reasonable within the realm of scientific possibility. Sometimes, however, genetic engineering can be so bizarre that it sounds like science fiction. For example, genetic engineers have been able to take genes from spiders and insert them into the DNA of goats. The result is healthy, normal goats. They just happen to produce the proteins for making spider silk in their milk. It sounds crazy, but scientists can milk a goat, harvest the proteins, and make spider silk. It can be made in much larger quantities than can be harvested directly from spiders. The silk has potential for a wide range of medical and industrial uses.

Genetics Problem

Choose a specific problem that potentially could be solved through genetic engineering. You may choose from the examples below or describe another problem:

- Areas that are densely populated have a lot of people to feed but not a lot of land for growing crops.
- Climate change is leading to more frequent periods of drought in several areas, making it more difficult to grow crops in these regions.
- Pesticides can be washed off plants when it rains, causing them to be ineffective.
- The cells in the pancreas of people with diabetes lose the ability to produce the protein insulin, which helps break down sugar.
- Fertilizers increase fossil fuel consumption and cause pollution from runoff.
- Milk prices are increasing as farmers struggle to keep up with the demand of a growing population.
- Fibers such as silk from silkworms are difficult to obtain in very large quantities.

Then, **explain** how this problem potentially could be solved through genetic engineering. This could be through the modification of the genes of a single organism or through the combination of genes from multiple organisms. These could include (among others):

- Corn
- Cows
- Fish
- Cotton
- Goats
- Trees (fruit or lumber)
- Wheat
- Spiders
- Rice
- Microbes

ELABORATE

Activity 14

Concept Review

Review: Modifying Organisms

Quick Code
ca8592s

Now that you have completed the objectives for this concept, **review** the core ideas you have learned. Record some of the core ideas below.

Core Ideas

Talk with a Group

Now, think about the video Searching for a Cause that you saw in Get Started. Discuss how what you've learned about modifying organisms can help you understand the search for the cause of dinosaur extinction.

EVALUATE

Solve Problems

Unit Project: Dinosaur De-extinction?

Quick Code
ca8595s

Could we bring back extinct animals? **Read** the text and **watch** the video to learn more.

Dinosaurs Once Roamed Earth

In the science fiction movie thriller *Jurassic Park*, scientists found preserved DNA from extinct dinosaurs and used it to bring extinct species back to life. This process, which is sometimes called de-extinction, has not yet been achieved, but many scientists are interested in the possibility. Over hundreds of millions of years,

Earth has experienced five mass extinctions in which a large percentage of living things disappeared. The most recent mass extinction occurred at the end of the Cretaceous period, about 65 million years ago. Today, human activities are credited with starting yet another mass extinction. When organisms go extinct, entire ecosystems can collapse. Some scientists are working on ways to restore ecosystems by restoring the extinct organisms that maintain them.

To accomplish this, they must find DNA of the extinct animals. DNA is present in all cells of living organisms. However, it is rarely preserved in fossils, particularly if there is no preserved soft tissue. Sometimes, small fragments of DNA have been discovered in frozen fossils. For example, scientists have found 24,000-year-old DNA of woolly mammoths in Siberia, and 800,000-year-old DNA of plants and insects in the Greenland ice cores. Obtaining 65-million-year-old DNA from dinosaur fossils is far less likely. Most of these fossils have become mineralized, and no cellular material remains.

Restoring Ecosystems with Extinct Species

SEP Obtaining, Evaluating, and Communicating Information

CCC Structure and Function

Theoretically, scientists should be able to use fragments of DNA of an extinct organism to modify a modern organism and create a new organism that functions as the extinct one did. Scientists may even be able to transplant complete frozen DNA into a modern living cell, which then divides to grow an exact replica of the extinct creature. Sound like science fiction? The reality of these techniques could be closer than you think.

Technical Challenges

Complete the statements below about technical challenges of de-extinction.

Bringing back extinct species requires ________________, which is usually not preserved because of ________________. This is why ________________ would be especially difficult to bring back.

Bringing Back Tasmanian Tigers

Tasmanian tigers are thought to have recently gone extinct. **Research** this animal and the following others: Tasmanian devil, Siberian tiger, African lion, gray wolf, and numbat. Then, **select** the most suitable modern animal to use to help bring back Tasmanian tigers using transgenics.

- ○ A. Tasmanian devil
- ○ B. Siberian tiger
- ○ C. African lion
- ○ D. Gray wolf
- ○ E. Numbat

Pros and Cons

Research the pros and cons of bringing woolly mammoths back to the Siberian tundra. **Describe** two pros and two cons in the space below.

UNIT
4
Monitoring
Biodiversity
Discovery
EDUCATION

Sensing Biodiversity from the Sky

Imagine using electromagnetic waves to survey the animals, plants, and even the chemistry of a tropical forest from an airplane. Scientists now have the tools to do this. How will this impact how we can manage and protect Earth's resources?

Quick Code
ca8751s

Guiding Questions

1. What are the characteristic properties and behaviors of waves?
2. What human activities harm Earth's biodiversity, and what human activities help to sustain global biodiversity?
3. How does communication technology encode information, and how can digital technologies be used to help sustain biodiversity?

Sensing Biodiversity from the Sky

Solve Problems

Unit Project: Bringing Back from the Brink

Quick Code
ca8750s

Thinking About Solutions

Could we bring back extinct animals?

You are part of a team of scientists interested in restoration programs for organisms, and how can they be improved upon. Before moving forward, **determine** a restoration program to focus on. **Consider** your ideas to improve it, the details of that idea, and what it might look like.

Park Ranger Holding Alligator Skin

Use the graphic organizer to sort your ideas.

My Idea

Details

Sketch

CONCEPT

4.1

Nature of Waves

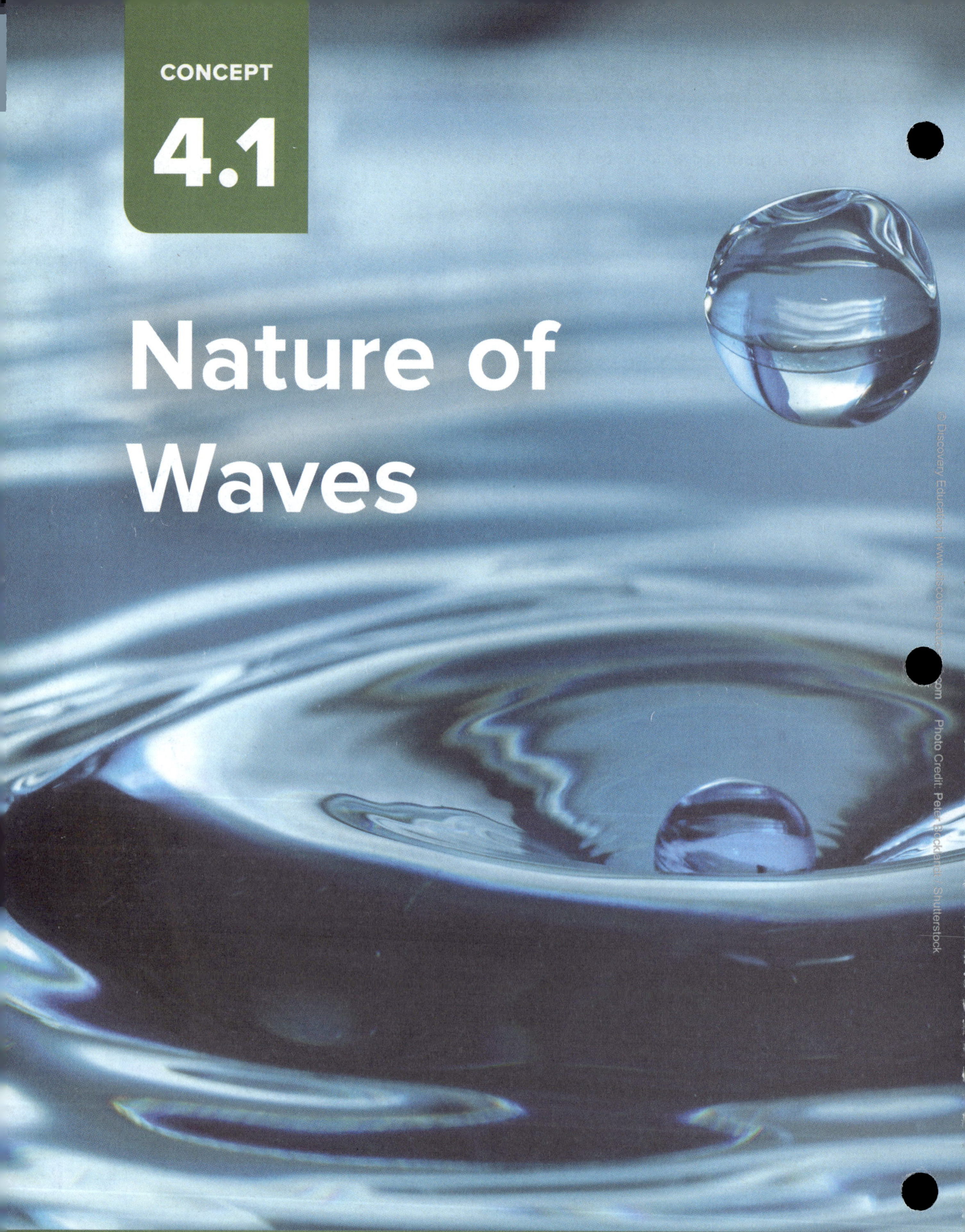

Student Objectives

By the end of this lesson:

- ☐ I can plan and conduct investigations to describe patterns in the characteristics of different types of waves.
- ☐ I can use mathematics and computational thinking to describe the relationships among wavelength, speed, and frequency of waves and to predict how changes in one variable affect other variables.
- ☐ I can use mathematical representations to construct an explanation of the relationship between amplitude and the energy in a wave.
- ☐ I can develop models that describe similarities and differences between mechanical and electromagnetic waves.
- ☐ I can argue from evidence about the advantages of digital versus analog signals.
- ☐ I can integrate qualitative scientific and technical information to support the claim that digital signals are more reliable than analog signals for encoding and transmitting information.
- ☐ I can obtain, evaluate, and communicate information about how waves can be used to transfer information.

Key Vocabulary

amplitude, energy, frequency, hertz, interference, light, longitudinal waves, sound, speed, transverse waves, vacuum, vibration, wave, wavelength

Quick Code
ca8753s

Activity 1

Can You Explain?

What are the similarities and differences between mechanical and electromagnetic waves?

Quick Code
ca8754s

Activity 2

Ask Questions

A Supernova

Quick Code ca8755s

Watch the following video, and then **answer** the questions that follow.

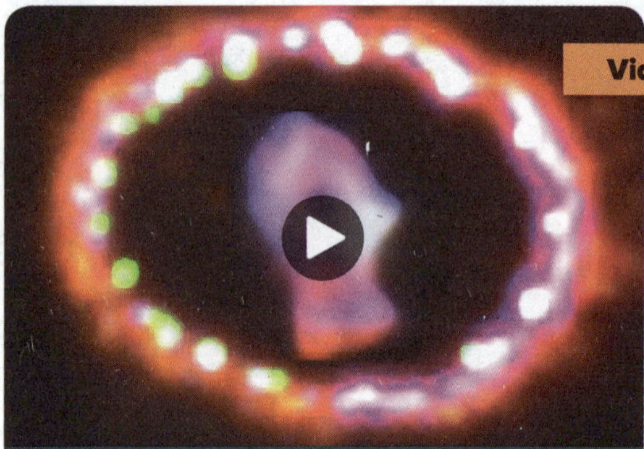

Let's Investigate a Supernova

Why can't we hear these explosions? Is it simply a matter of distance or is something else preventing the noise from getting to our ears?

What other questions do you have about supernovas?

ENGAGE

Activity 3

Analyze

Explosion in Space

Quick Code
ca8756s

As you read this text, **identify** and **record** the differences between the nature and behavior of sound and electromagnetic waves.

Explosion in Space

Do you enjoy science fiction movies about space? Have you noticed that, in most cases, such movies show moving spaceships with roaring engines and noisy explosions? Why are those scenes inaccurate? You might have watched an accurate video that shows explosions in outer space with no accompanying **sound**. It seems strange since sound is part of our everyday lives. On Earth, people hear all sorts of different sounds all the time, from ringing phones to passing cars to chirping birds. Why is there no sound in space?

Both sound and **light** are forms of **energy** that travel as waves. Sound travels through matter—it is a matter **wave**. Light, on the other hand, is a type of electromagnetic radiation. As its name suggests, electromagnetic radiation is made up of electricity and magnetism. It does not need matter to travel. This means it can travel through a **vacuum**, which explains why we can see an explosion in space but cannot hear it.

In this concept, you will learn more about matter waves like sound and about electromagnetic waves like visible light. You will discover their similarities and differences, as well as some of the ways they are used in modern technologies.

SEP **Engaging in Argument from Evidence**

ENGAGE

Use a Venn diagram to **record** the differences and similarities between the two types of waves (sound and electromagnetic).

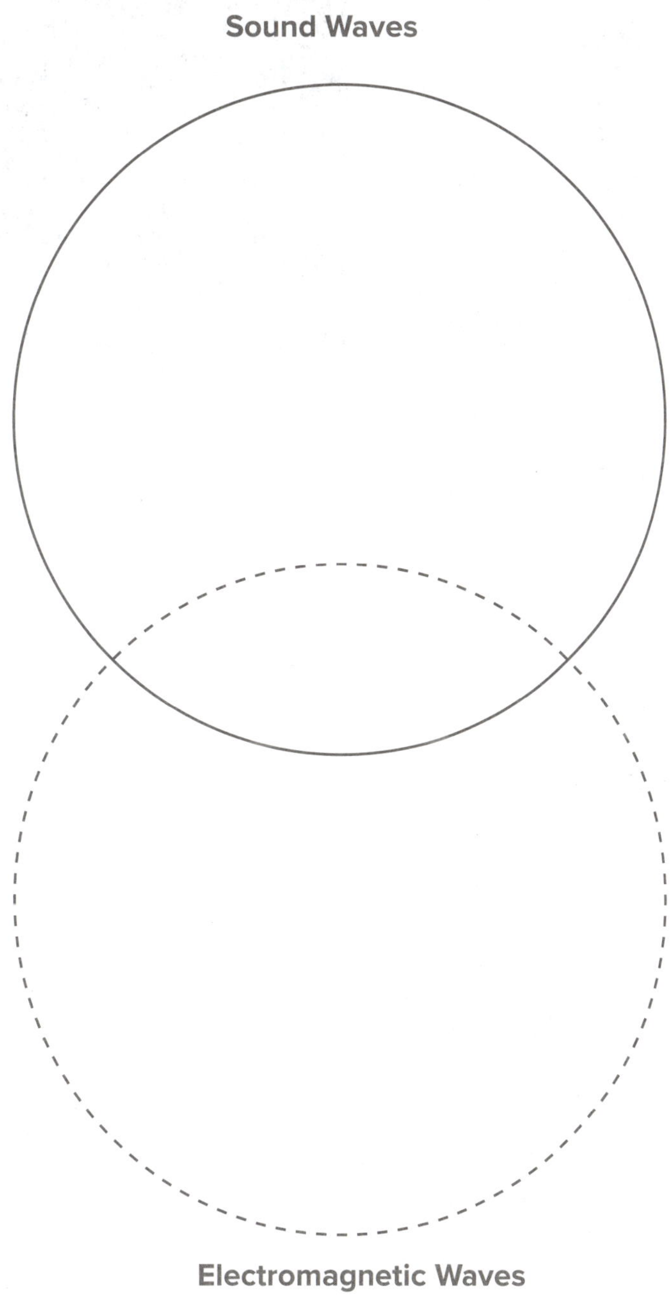

Look at the images and **discuss** the following questions with your group.

Shout or Scream

Lightning

What happens when you shout to a friend?

How does the message get to that person? Does it only travel to them or can it be heard elsewhere?

ENGAGE

What about light from a bulb? How does it behave when it leaves the bulb?

What happens to both light and sound the farther you get from the source? When lightning flashes, is thunder produced at the same time? Do both types of waves travel at a similar speed?

Activity 4

Evaluate

Quick Code
ca8757s

What Do You Already Know About the Nature of Waves?

What Are Waves?

The following statements describe waves. **Select** the correct term from the word bank to complete each statement.

crest	trough	rarefaction	compression
parallel	wavelength	amplitude	perpendicular

A longitudinal wave oscillates ____________________ to the direction in which it travels.

A transverse wave oscillates ____________________ to the direction in which it travels.

The ____________________ is a point on a wave where the medium is farthest above its resting point.

The distance between two troughs or two crests is the ____________________ of the wave.

The maximum distance that the wave moves from its resting position is the ____________________ of the wave.

ENGAGE

Now, **draw** a sketch of a transverse wave and label it with the various properties of waves.

Ordering Colors of Light

What is the order of the colors in the visible spectrum, from longest wavelength to shortest wavelength?

____

____

____

Where and why do you commonly see these colors shown grouped together in order?

____

____

____

What Are Waves?

Activity 5

Investigate

Quick Code
ca8758s

Hands-On Investigation: The Ripple Effect

In this investigation, you will investigate the transmission of waves through a fluid.

Predict

What do you predict will be the best way to measure waves generated by dropping an object into water?

SEP **Engaging in Argument from Evidence**

What materials do you need? (per group)

- Plastic pan
- Golf balls
- Disposable trays
- Toothpicks
- Metric ruler
- Masking tape
- String
- Small weights (for anchors)
- Stopwatch
- Salt
- Water

HANDS-ON INVESTIGATION

EXPLORE

Procedure

1. Wrap a piece of tape around a toothpick to create a small flag.
2. Using the flag, poke a hole through a small piece of polystyrene foam or cork and insert the flag near the hole.
3. Thread a piece of string through the hole and tie it off so it does not slip out.
4. Attach a small object to the other end of the string using tape. Follow this procedure to create 4–6 of these "buoys."
5. Place these buoys at different places in a basin of water.
6. Then, design an experiment in which you drop objects into the water and observe how waves travel outward. You can measure the time a wave passes each buoy by observing when the buoy will bob up and down. You can explore different variables to see how the size and speed of the waves change. Include a question to investigate, your investigative procedure, and a way to record and analyze results.

Design an experiment that will help you determine how the makeup of the water affects the waves produced. Describe your experiment.

What objective data could you gather in this experiment?

Reflect

How does the mass of an object affect the speed and size of the wave it creates? How does the height from which the object was dropped affect these aspects of the wave?

How does changing the makeup of the water by adding salt change the waves produced?

How does your experiment demonstrate the potential and kinetic energy conversions between the object, water, and buoys?

EXPLORE

Activity 6

Observe

Ripples

Quick Code
ca8759s

Look at the image and **answer** the questions that follow.

Ripples

CCC Patterns

How does energy flow in the image?

What evidence did you collect in the Hands-On Investigation that supports your explanation? How does the evidence support it?

EXPLORE

Activity 7

Analyze

Introduction to Waves

Quick Code
ca8760s

Watch the following video and **read** the text. **Read** the following text and **underline** any information you can use as evidence to support your initial ideas about the similarities and differences between mechanical and electromagnetic waves.

Introduction to Waves

What Are Waves?

What are waves, how do they behave, and what do they do? Waves are disturbances in space or matter that carry energy from one place to another. Some waves are visible; others are not.

All waves start from an initial **vibration**, or oscillation, that provides some energy. A vibrating guitar string produces sound waves. Vibrating atoms in a light bulb produce light waves, and vibrations in Earth produce seismic waves. All waves start with a source. As the wave moves away from its source, it transmits the energy from one place to another. How does it transmit this energy?

Consider the motion of waves in water. When water is disturbed, waves move out away from the disturbance. The surface of the water moves up and down and energy from the wave radiates from the source.

Energy released at the source causes the particles in a body of water such as a lake to move up and down and side to side.

These particles oscillate, but their average position remains the same. In a lake, this motion is typically caused by the wind or a passing boat. As particles oscillate, they disturb the particles next to them. This causes those particles to oscillate and the oscillation travels through the water. It is said that the wave propagates through the water.

To observers on the shore, it may look as though the water in the lake is moving toward them. However, only the energy of water particle oscillations is transmitted toward the shore. The water particles are not actually moving closer to shore. As the waves break on the shore, only the energy of the particles in them is transferred.

EXPLORE

Revisit your explanations from the Ripples image. **Refine** your explanation using evidence from the video and text.

____

____

____

____

____

____

____

____

____

____

____

Activity 8
Analyze

Quick Code
ca8761s

Transverse and Longitudinal Waves

Read the text and **highlight** information that supports your explanation for the behavior of waves in water.

Transverse and Longitudinal Waves

When particles oscillate up and down relative to the direction that the wave propagates, the wave is a transverse wave. This is the most familiar type of wave. However, sometimes the particles in a wave oscillate in the same direction that the wave propagates. These waves are called **longitudinal waves**. Sound waves are an example of longitudinal waves. They are also called pressure waves or compression waves.

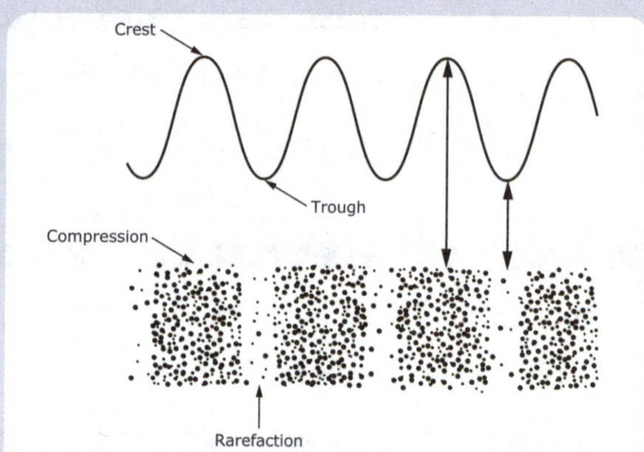

Transverse Versus Longitudinal Waves

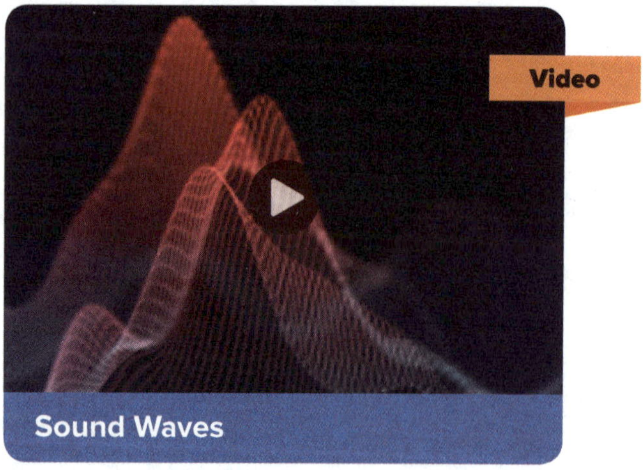

Sound Waves

CCC Patterns

EXPLORE

Now, **use a** Venn Diagram to **record** the differences and similarities between the two types of waves (transverse and longitudinal).

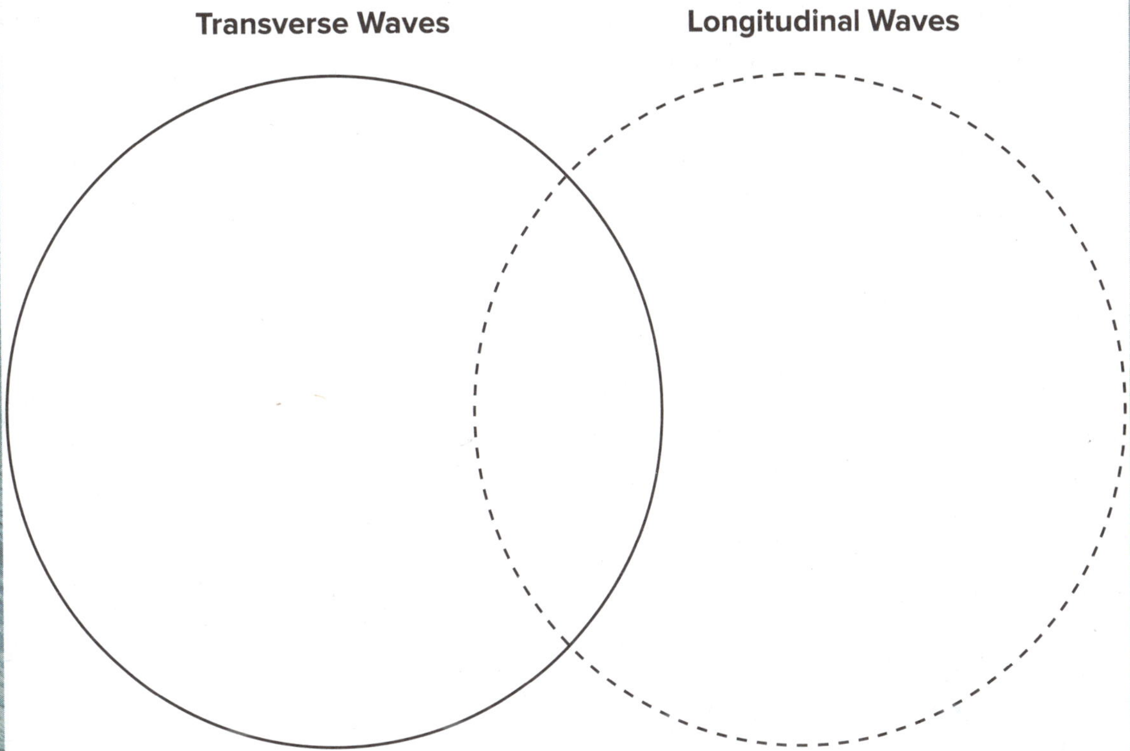

Activity 9

Investigate

Quick Code
ca8762s

Hands-On Investigation: Transverse and Longitudinal Waves

In this investigation, you will work in pairs to investigate how two models can be helpful in describing and comparing the characteristics of transverse waves and longitudinal waves.

Predict

Do you predict you will be able to use the same models to describe transverse and longitudinal waves? Explain your answer.

SEP Engaging in Argument from Evidence

What materials do you need? (per group)

- Rope
- Metal spring toy

HANDS-ON INVESTIGATION

Procedure

1. Sit on the floor about 2 meters away from your partner. One of you will create the wave using one end of the rope and one of you will hold the other end of the rope steady.
2. The student who is creating the wave with the rope should move their end at a steady pace from side to side while the other student holds their end steady.
3. Observe the shape of the wave generated by this motion.
4. Then, stop moving the rope. Both students should let go of the rope, being careful not to change its shape.
5. Draw the shape of the rope.

6. Then, repeat the movement of the rope while varying how vigorously it is moved from side to side.
7. Observe how the wave changes as the motion is varied.

____

____

____

____

____

____

8. Next, swap roles with your partner as you use the spring. Sit 2 meters apart on the floor. Each of you should take one end of the spring in your hand.
9. Pull the spring so there is a slight pull from each student.
10. The student who is creating the wave should push their end of the spring toward their partner and then quickly pull it away.
11. Observe the shape of the wave generated by this motion.
12. Repeat this motion but vary how far the end of the spring is pushed before pulling it back. Observe the effects. Make a drawing of the shape of the spring.

Reflect

What similarities did you see in the waves that you generated using the rope and the spring? What differences did you see?

What changed when you varied how vigorously you moved the rope or spring? What does this tell you?

What is an example of a real-world wave modeled by using the rope? What is an example of a real-world wave modeled by using the spring?

EXPLORE

What Are the Characteristics of Waves?

Activity 10

Analyze

Characteristics of Waves

Quick Code
ca8763s

Read the text and **underline** sections that describe each of the vocabulary words in the table.

Characteristics of Waves

One characteristic of a **wave** is its **wavelength.** This is the distance over which the wave completes one oscillation. To determine this distance, locate the peak of the wave. Then, observe the next oscillation of the wave and find the location of the next peak. The distance between these two peaks is the wavelength. Since wavelength is a measure of distance, it is measured in meters (m) and is usually denoted by the Greek symbol λ (lambda).

This diagram shows two complete cycles of a wave. Notice that each cycle has the same wavelength. What points does a complete wave pass through?

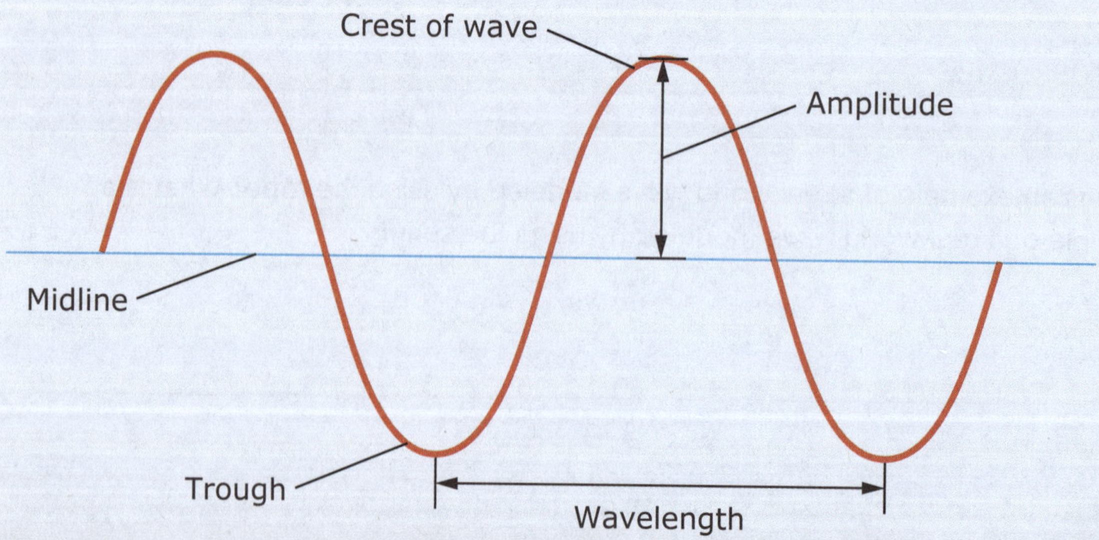

The period of a wave is similar to its wavelength. The wave period is the time it takes for a wave to complete one full oscillation. The period is measured in seconds and is represented by the symbol T.

The **frequency** of a wave is the number of oscillations that a wave makes each second. Frequency is measured in cycles per second, which has units of 1/s. This unit, 1/s, is referred to as **hertz** $1T$ and $T = 1f$.

The **amplitude** of a transverse wave is the height of the wave. Since amplitude is a measure of height, it is expressed in meters and is typically denoted by the symbol A. The amplitude of a wave is related to its **energy.** Think of yourself at the beach. Which wave will knock you over—a small wave or a big one? Which has the most energy?

The wave **speed** is the speed at which the wave propagates through matter or space. Wave speed has units of meters per second, and it is typically expressed by the symbol v. The speed at which a wave travels depends upon the type of wave and the nature of the medium that it is passing through. For example, electromagnetic waves travel fastest in a **vacuum**, but slower (and sometimes not at all) through matter. All other waves require matter for their transmission. For this reason, they are called mechanical waves. **Sound**, seismic, and water waves are examples of mechanical waves.

EXPLORE

Watch the following video. **Record** your own definitions for each term in the table.

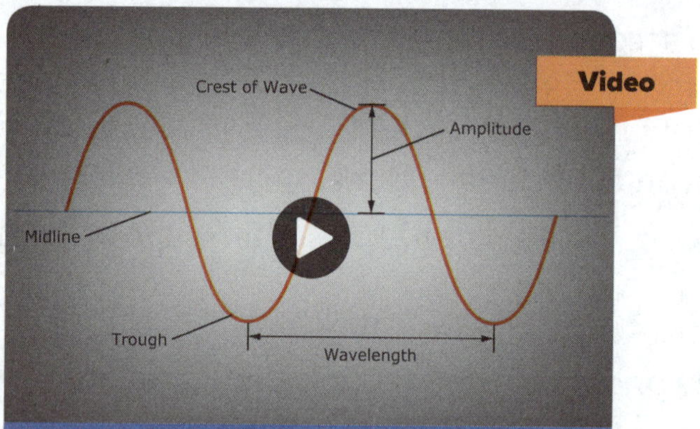

Relationship between the Parts of a Wave and Its Energy

Term	Definition
Crest of Wave	
Midline	
Trough	
Amplitude	
Wavelength	

Activity 11

Observe

Describing Waves

Quick Code
ca8766s

Go online to familiarize yourself with different wave properties. **Change** combinations of these parameters and **observe** the results of these changes. **Record** your results in a table, graph, or diagram.

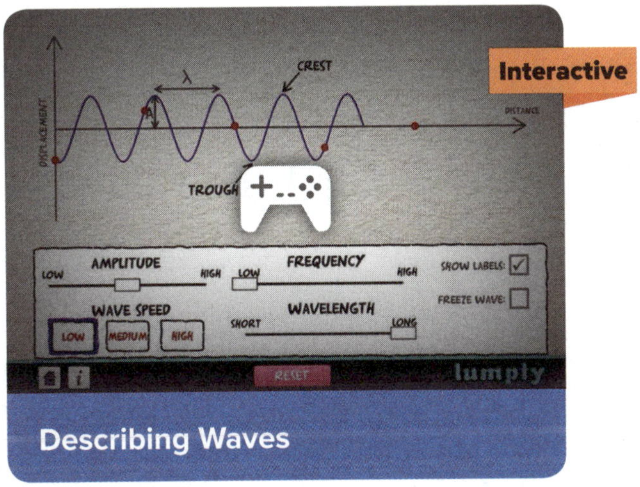

Describing Waves

Record your results.

SEP Planning and Carrying Out Investigations

CCC Patterns

EXPLORE

Activity 12

Reason

Quick Code
ca8767s

Measuring the Parts of a Wave

In this activity, you will study the measurements of parts of several waves. You will use these measurements to calculate properties of the waves and compare them.

What materials do you need? (per group)

- Metric ruler

Procedure

1. Look at the data sheet.
2. Use these measurements of waves to analyze the properties of the waves and compare them to one another.

SEP Engaging in Argument from Evidence

CCC Patterns

Wave Number	Wavelength (cm)	Frequency (wave/second)	Amplitude (cm)
1			
2			
3			
4			
5			

EXPLORE

Examine the graphs of waves that follow. How do the waves compare? You may include drawings.

Graph 1

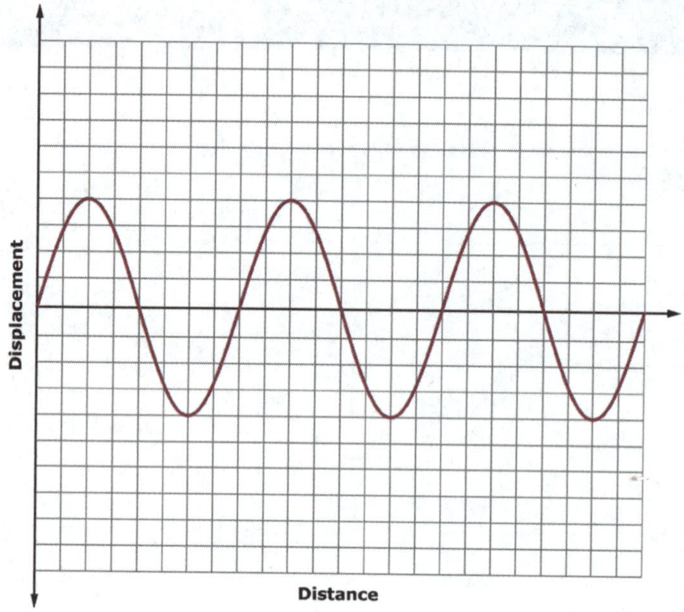

Graph 2

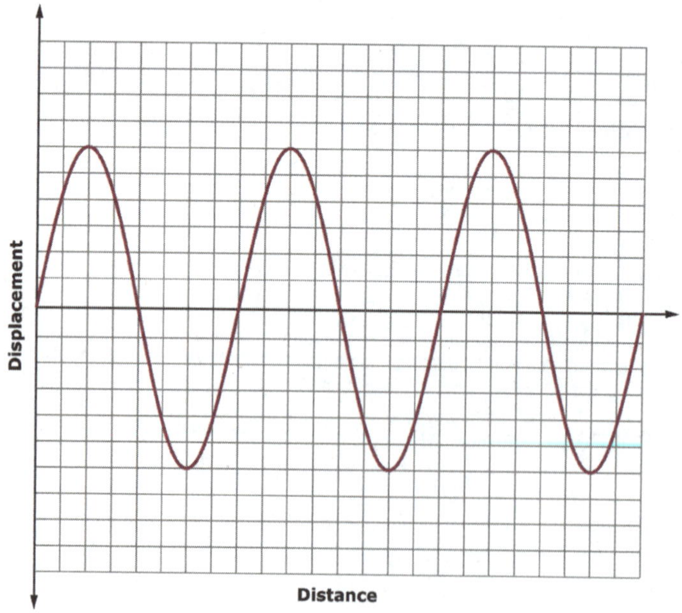

Graph 3

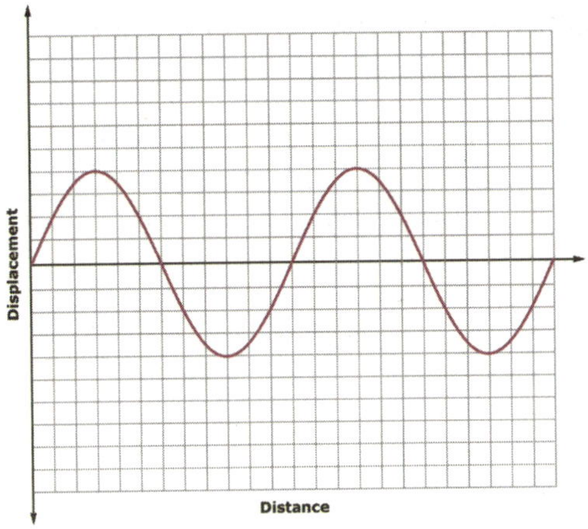

Graph 4

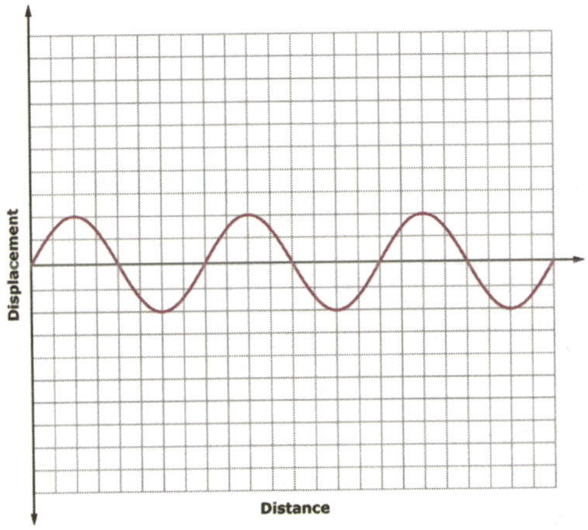

Graph 5

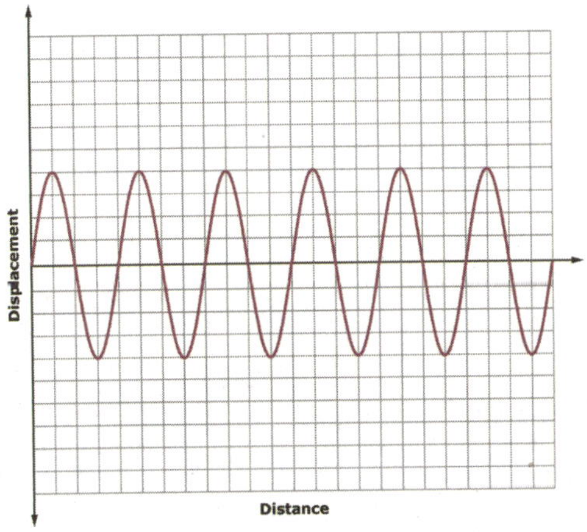

EXPLORE

Reflect

Which wave has the most energy? How do you know?

Which wave has the highest frequency? How do you know?

How would the frequency of the waves change if their speeds were doubled?

How would the energy of the waves change if their wavelengths were doubled?

EXPLORE

Activity 13

Evaluate

Properties of Waves

Quick Code
ca8768s

The figure represents two different waves. The top diagram shows a wave that is measured over distance, whereas the bottom diagram shows a wave that is measured over time. The letters indicate various properties of the waves. **Write** whether each statement is true, false, or neither true nor false.

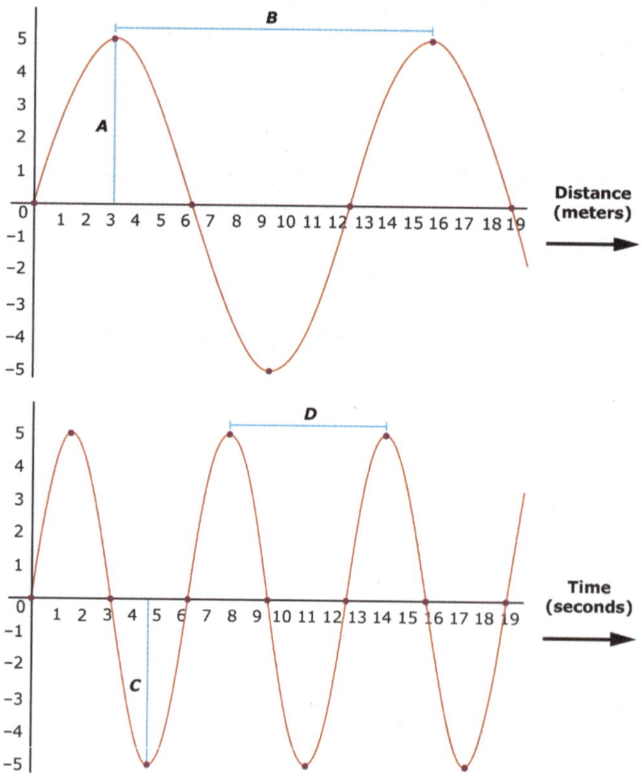

Line A represents the wave's amplitude.
Lines B and D represent the waves' wavelengths.
Line C represents the wave's amplitude.
Line D represents the wave's period.
Only one of the waves could represent a light wave.
The wavelengths can only be measured in the top diagram.
The wave in the top diagram is moving more slowly than the wave in the bottom diagram.

SEP Analyzing and Interpreting Data

What Are Mechanical and Electromagnetic Waves?

Activity 14

Observe

Quick Code
ca8769s

EXPLORE

Properties of Light

Watch the following video and **answer** the questions.

Video

Properties of Light

What is the electromagnetic spectrum?

____

____

____

Which property of a light wave determines where it is on the spectrum?

____

____

Look at the image.

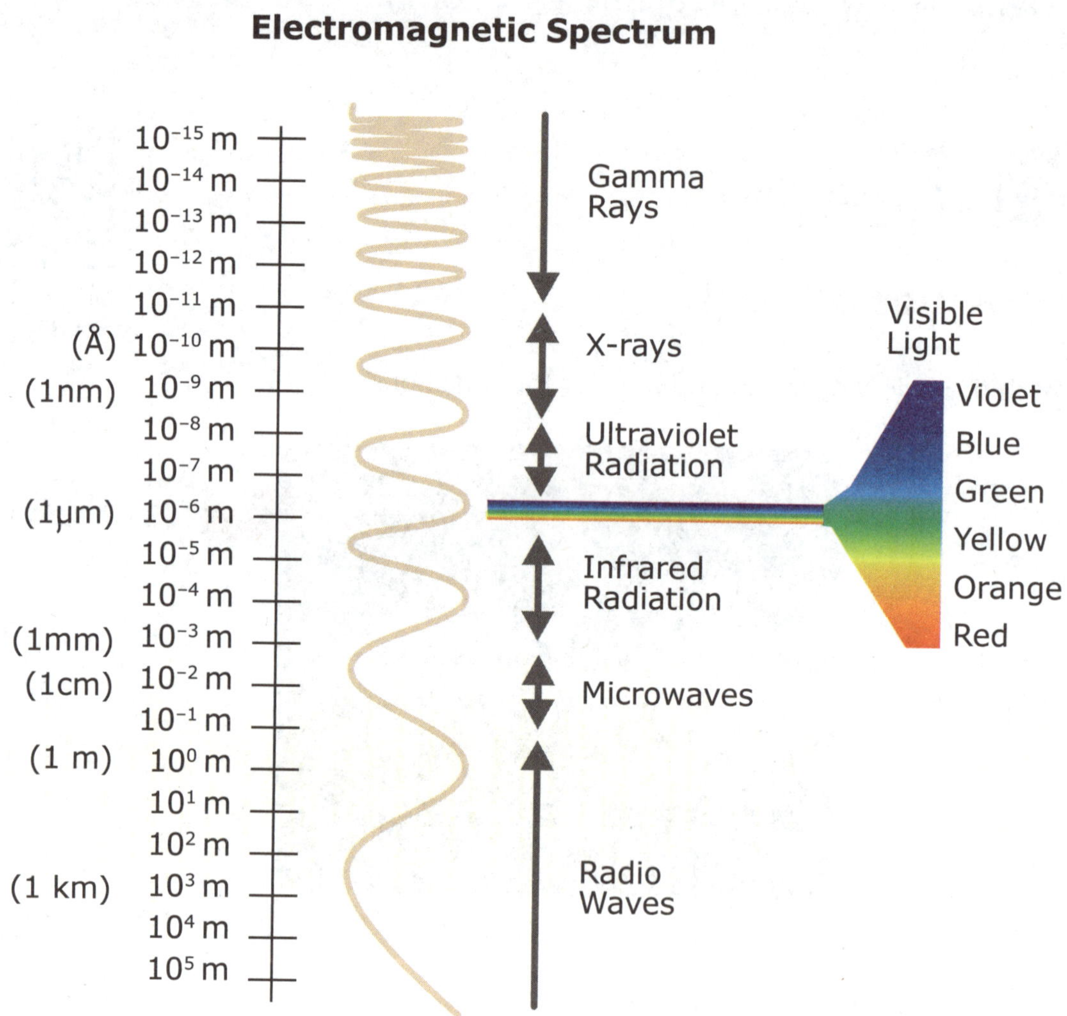

Now, **record** the characteristics of different types of electromagnetic waves and examples of their real-world applications in the chart.

Characteristics of Electromagnetic Radiation	Real-World Application

EXPLORE

Activity 15
Observe

Quick Code
ca8771s

On Your Wavelength

Go online and follow the procedure given in the graphic organizer to explore visible light. **Answer** the questions in the organizer.

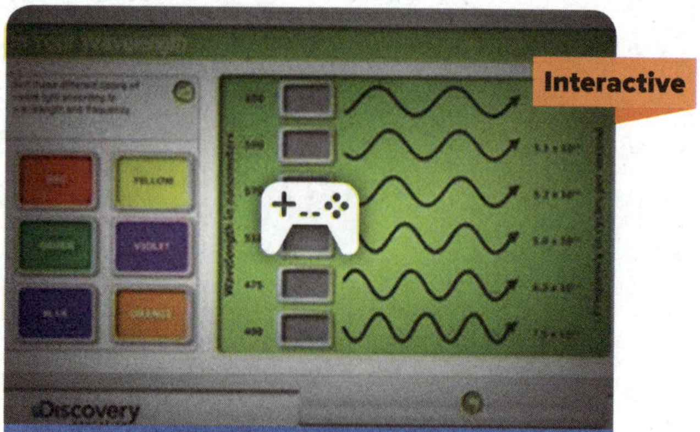

On Your Wavelength

CCC Patterns

EXPLORE

Overview

In this exploration, you will sort the different colors of visible light according to wavelength and frequency, and then discover visible light's place in the electromagnetic spectrum.

Questions

1. Which color in the visible spectrum has the most energy? ______________________

2. Why isn't the sky orange? ______________________

____

____

3. What is the correlation between the length of a light wave and its frequency?

____

How to Use This Exploration

1. **Read** the Introduction and **click** the Continue button.
2. **Identify** the exploration components. **Note** that one nanometer is equal to one trillionth of a kilometer.
3. **Recall** the order of colors in a rainbow. **Drag** each color to its position and **read** the explanation.
4. **Continue** placing all the colors.
5. **Read** the Congratulations text and **click** on the Next button.
6. **Read** and **take notes** on the outcome animation information.

Data

____

____

____

Activity 16

Reason

Investigating Spectra

Quick Code
ca8772s

In this activity, you will use spectroscopes to compare the spectra from different sources of light, such as sunlight, fluorescent light, and incandescent light.

What materials do you need? (per group)

- Spectroscope
- Spectral discharge tubes with power supplies (optional but recommended)
- Flashlight
- Batteries, size D
- Fluorescent light (such as the overhead lighting in many classrooms)
- Colored pencils

Procedure

1. Turn off the overhead lights and open the window shades so that sunlight streams into the room.
2. Aim the light-gathering end of the spectroscope at the sky.
3. Then, draw the image that the spectroscope produces and label it "sunlight." NOTE: Never look directly at the sun or aim a spectroscope directly at the sun.

4. Turn on the overhead lights and close the window shades.

5. Aim the light-gathering end of the spectroscope at the overhead lights, and then draw the image from the spectroscope. Label it "overhead lights."

6. Turn off the overhead lights and turn on a flashlight.

7. Aim the light-gathering end of the spectroscope at the flashlight, and then draw the image from the spectroscope. Label it "flashlight."

EXPLORE

Reflect

Compare the spectra from different light sources. How are they similar? How are they different?

Why do you think spectroscopes are useful for studying the sun and other stars?

If students or scientists in locations around Earth used a spectroscope to study sunlight, do you think the spectra they observe would be the same? Explain your prediction.

If astronauts ever visited Mars and used a spectroscope to study sunlight there, do you think the spectra they obtain would be the same as what you observed today? Explain your prediction.

EXPLORE

How Can Waves Be Used to Transfer Information?

Activity 17

Analyze

Using Waves to Transfer Information

Quick Code
ca8773s

Brainstorm all the methods of communication you can think of.

____

____

____

Can you identify the methods that use waves to communicate?

____

____

Can you identify forms of communication that use electromagnetic waves?

____

____

____

SEP **Engaging in Argument from Evidence**

Answer the questions. Then, as you read the text that follows, **identify** and **highlight** the advantages of using electromagnetic waves over other waves used in communication.

EXPLORE

Using Waves to Transfer Information

Because waves travel through space and matter, they can be used to carry information. Humans commonly manipulate sound and electromagnetic waves for this purpose. Sound is used for speech, but sound travels relatively slowly (about 343 meters per second in air), and even at high amplitudes (volume) it rarely travels distances of more than a few kilometers. Electromagnetic waves travel about a million times faster through air (3×10^8 meters per second) and can travel through the vacuum of space and some solids (such as glass). This makes them useful as an almost instantaneous form of communication over very large distances, from radio to fiber optics.

Electromagnetic waves can easily be made by using electricity to excite electrons in atoms. For example, we can use sound to generate electrical energy, which in turn can be used to send out electromagnetic waves. The first step of this sequence can be accomplished using a device called a microphone. Microphones, like the ones found in cell phones, convert vibrations in the air (sound) into vibrations in a thin sheet called a diaphragm. This is a thin piece of material (such as paper, plastic, or metal) found in the head of the microphone. When the diaphragm vibrates, other components in the microphone convert these vibrations into an electrical current, which becomes the audio signal.

Watch the following video and **answer** the question that follows.

Communicating with Light

Why are optical fibers replacing copper wires as the preferred method of transmitting information?

Activity 18

Reason

Going Digital

Quick Code
ca8774s

In this activity, you will digitize an image using a grid.

What materials do you need? (per group)

- Black marker

Procedure

Fill in every black area from the image onto the grid. Every white area on the image should be left blank on the grid. Any square that is filled in must be filled in completely.

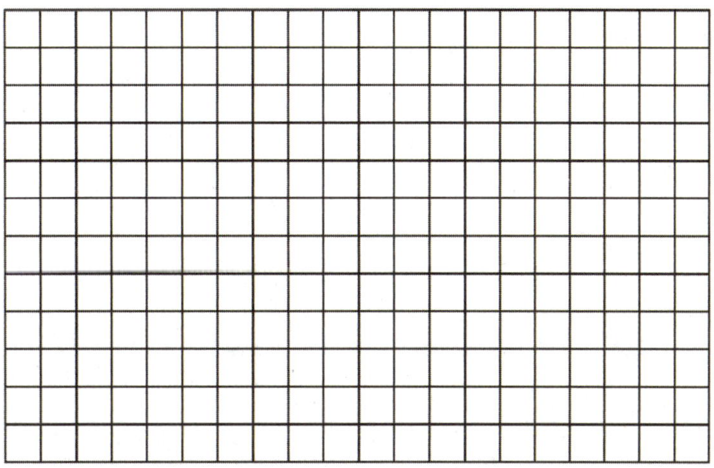

CCC **Patterns**

Reflect

How is the digital image like the original? How is it different?

What would it take to make your digital image more closely resemble the original?

What does this tell you about resolution and image file size?

Activity 19

Analyze

Quick Code
ca8775s

Analog versus Digital

Think about how phones enable us to communicate with one another.

How do you think digital phones (cell phones) transmit signals?

____

____

____

EXPLORE

Look at the images of an analog signal of flute music and a comparison of digital and analog signals. Then, **read** the text that follows. As you read, **highlight** all the advantages digital signals have over analog ones.

An Analog Signal of Flute Music

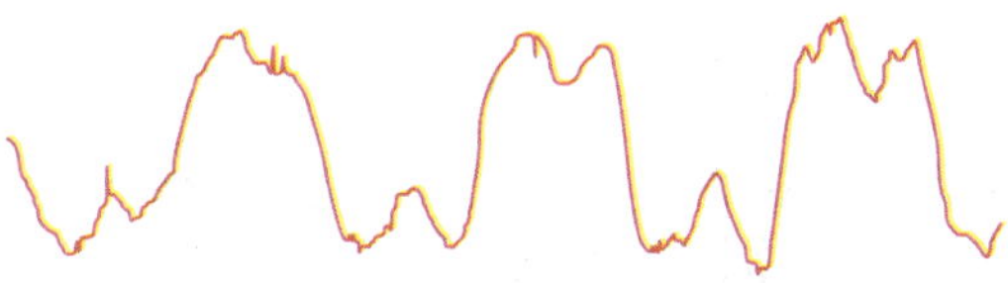

Digital versus Analog Signals

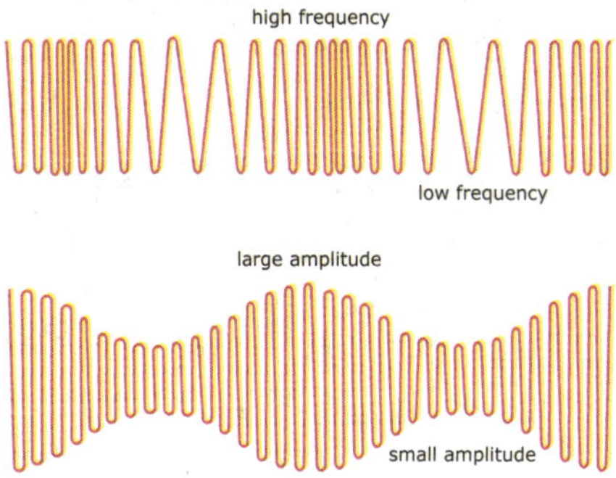

CCC Patterns

Analog versus Digital

Most natural signals are analog signals. While an analog signal is composed of an infinite possibility of values, a digital signal is not infinite. It uses discrete values of wave height. Morse code is an example of a digital signal. The dots and dashes are transformed into discrete values when transmitted. The message can then be decoded by a person listening to or watching the transmission. The advent of transistors and computers has allowed scientists and engineers to digitize analog signals (simply called digital signals). They do this by assigning a specific digital value to the height of a wave at small intervals along the wave.

Digital signals have advantages over analog signals. These signals may easily be stored and manipulated by computers. They can also be more easily reproduced than an analog signal. Perhaps most importantly, digital signals are less affected by **interference** that they encounter when they are being transmitted.

While a signal is moving through a wire or space, it loses energy and many factors can interfere with it, making the received signal different than the transmitted one. Engineers call this noise. In an analog signal, every alteration in the shape of the wave by noise is "heard" on the receiving end. In a digital signal, if the change in shape is not too great, the signal will stay the same because the system knows the discrete wave height that should be present there and can reproduce it. Today's popular media (Internet, broadcast television, some radio, cable television) are all based on digital signals.

What would your life look like if you had to use only analog signals?

EXPLORE

Activity 20

Evaluate

Quick Code
ca8776s

Analog and Digital Signals and Two Types of Signal

Analog and Digital Signals

Compare and **contrast** what happens when someone uses an analog signal versus a digital signal to communicate in a particular situation. **Use** a real-world scenario for the situation.

Two Types of Signal

Fill in the blanks for each sentence with "less" or "more" to complete the sentences.

In general,

- A signal that is continuously varying is ________________ reliable than a signal with limited possible values.
- A signal stored on a vinyl record is ________________ easily copied than a signal stored in a computer.
- A signal transmitted by the human voice loses ________________ energy as it travels than a signal transmitted by a fiber optic cable.
- A clock that uses hands to tell the time is ________________ precise than a clock that uses numbers.

EXPLORE

Activity 21

Solve Problems

Quick Code ca8778s

Digital Communication

Analog signals, such as standard radio, are subject to noise and interference. Digital signals do not suffer from these problems. For this reason, digital communication is becoming the method of choice for sending large amounts of information across distances. Even television and radio signals are increasingly sent as digital rather than analog signals.

Digital Communication

All digital communication uses code to send information. A digital code usually consists of a binary code. In this code, the signal is either "on" or "off." This is represented with numbers: 1 for "on," and 0 for "off."

Plan a method to send a digital code bidirectionally (meaning you can send and receive code from either end) by flashlight or wire. You will write your code in the next step. For your plan, **state** your objective, that is, what you intend to accomplish. **Provide** a materials list, a detailed description of safety precautions you will take, and a step-by-step procedure. Be sure to include what data you will collect.

SEP Constructing Explanations and Designing Solutions

SEP Developing and Using Models

What is your objective? Are there any safety precautions?

Describe your procedure.

What materials will you need?

Create your cypher, that is, your key to encoding or decoding a message. **Include** your cypher here.

EXPLORE

Carry out your plan using the cipher you designed. **Include** your results here and **analyze** them. In your analysis, **discuss** what worked well and what did not work well.

Based on your results, **revise** your plan. **Make** adjustments to your cipher, your procedure, or both. **Carry out** your revised plan and **provide** an analysis of the results. Be sure to explain whether your revisions improved your results. **Compare** your designs with those of other teams. Additionally, **explain** why using a code such as yours is better than using analog signals, such as your voice.

Use the graphic organizer to **record** the strengths and weaknesses of your initial design.

Strengths	Weaknesses

EXPLORE

Activity 22

Record Evidence

A Supernova

Quick Code
ca8777s

As you worked through this lesson, you investigated and gathered evidence about the nature of waves. Now, take another look at the video Supernova, which you first saw in Engage.

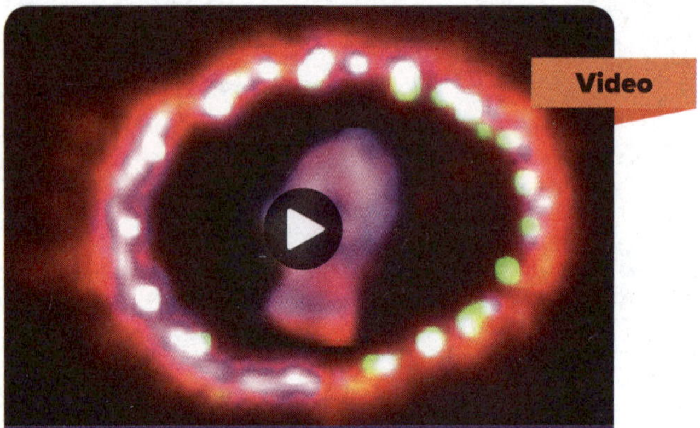

Let's Investigate a Supernova

How has your understanding of the Supernova video changed?

Read the Can You Explain? question from the beginning of this lesson.

Can You Explain?

What are the similarities and differences between mechanical and electromagnetic waves?

Use your new understanding of the Supernova video to write a scientific explanation answering a question.

1. Choose a question. You may choose to answer the Can You Explain? question or one of the questions you wrote at the start of this lesson.
2. Use the following information to help you write your explanation.

Construct an explanation that uses one or more models or representations as evidence. Models can include graphs, charts, drawings, photographs, or other types of graphics. Prior to writing your scientific explanation, think about what sort of models would best serve as evidence to support your claim. You can either draw your representations or print them out and attach them to the explanation.

Now, create a scientific explanation to answer the Can You Explain? question.

SEP Constructing Explanations and Designing Solutions

EXPLAIN

CONCEPT 4.1

What are the similarities and differences between mechanical and electromagnetic waves?

EXPLAIN

STEM in Action

Activity 23

Analyze

Engineering a Television Image

Quick Code
ca8779s

Read the following text. Then, **complete** the activities that follow.

Engineering a Television Image

Today's television screens can produce very detailed images. Can you believe that all of these images are produced from just three primary colors? The technology that produces images on television, cameras, and computer screen has evolved significantly over time. However, it all relies on constructing larger images from a combination of three colors. Let's explore how television engineers apply their knowledge of light and color to produce detailed television images.

Inventing the Television

Color television was first introduced to U.S. audiences in the 1950s. The first color TVs were cathode-ray tube television sets. A cathode-ray tube television contains three electron guns at the back of the television. Each electron gun shoots a beam of electrons at tiny dots that each glow a different color when hit by the electrons.

The three colors of light that produce all of the images on a screen are red, blue, and green. Every color we see is a combination of these three primary colors. This is known as additive color mixing. This may surprise you if you have ever mixed together the primary colors red, blue, and yellow when painting. Additive color mixing on a television screen works differently than the color mixing we see when we mix paints.

Recall that white light contains all colors of light. When we use a yellow paint on paper, the paper absorbs, or subtracts, all colors of white light except yellow. We see the color yellow because yellow light is the only color reflected from the paper. When we mix yellow and blue paint together to make green, it is called subtractive color mixing because all of the wavelengths of light are absorbed except for green.

ELABORATE

Engineering a Television Image *cont'd*

This process is different in a television. The display starts from darkness instead of white light. The three colors are added together, not subtracted, in dots on the screen to produce the range of colors we see.

Newer televisions no longer fire electrons to make dots glow. In liquid crystal display (LCD) televisions, a light is used at the back of the screen. When the light passes through liquid crystals on the screen, it turns tiny dots containing one of the three primary colors on and off. Engineers must understand how color combines to produce the pictures we see.

Color Mixing in LCD Pixels

An LCD television screen uses additive color mixing to create the following very simple image in 9 pixels. Determine which primary colors light in each LCD pixel to form the colors on this screen. **Write** one or more of the primary colors from the word bank on each square.

red	blue	green

ELABORATE

Activity 24

Concept Review

Review: Nature of Waves

Quick Code
ca8780s

Now that you have completed the objectives for this concept, **review** the core ideas you have learned. Record some of the core ideas below.

Core Ideas

Talk with a Group

Now, think about the video Sensing Biodiversity from the Sky, which you saw in Get Started. Discuss how what you have learned about the nature of waves can help you understand how the airborne observatory monitors life below.

EVALUATE

CONCEPT

4.2

Waves and Matter

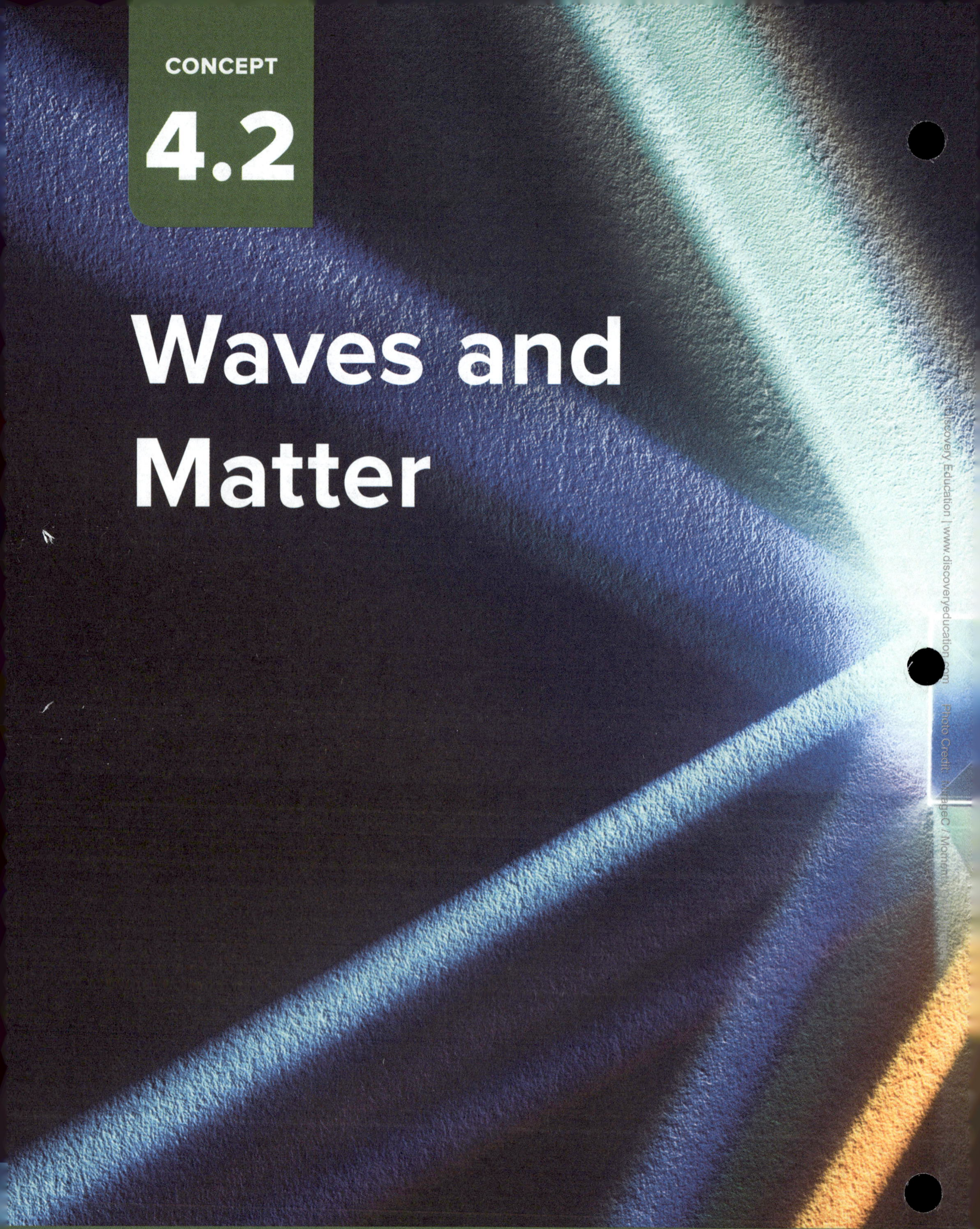

Student Objectives

By the end of this lesson:

- ☐ I can develop models that describe how waves are reflected, absorbed, or transmitted through various media.
- ☐ I can analyze and interpret data to describe and predict how the media through which waves move and the matter that waves encounter cause changes in wave behavior.
- ☐ I can evaluate competing designs for a sound-reflecting or sound-absorbing surface using a systematic process to determine how well they meet the required criteria.
- ☐ I can use mathematical representations and computational thinking to identify patterns in the way light reflects off a surface.
- ☐ I can plan and conduct an investigation to graphically represent, explain, and model the relative reflection and absorption of light from different colored objects.
- ☐ I can develop models that describe and predict how waves travel around objects on small and large scales.

Key Vocabulary

air, diffraction, energy, frequency, infrared, light, matter, reflect, refraction, sound, translucent, transparent, wavelength

Quick Code
ca8782s

Activity 1

Can You Explain?

What happens to a ray of light as it is reflected off a mirror and into a glass of water?

Quick Code
ca8783s

© Discovery Education | www.discoveryeducation.com • Photo Credit: Walter DuRant / Shutterstock

Activity 2

Ask Questions

Laser Art

Quick Code
ca8785s

Watch the following video, and then **answer** the questions that follow.

Video

Let's Investigate Laser Art

ENGAGE

Why are lasers different from "normal" light?

How can these properties explain the way they interact with matter?

What questions do you have about lasers?

Activity 3

Analyze

Using Laser Light

Quick Code
ca8786s

Read the following text about laser light. As you read, **complete** a T-Chart with two columns that compares the properties and uses of light from a normal light bulb with the light emitted by a laser. **Include** similarities as well as differences.

Using Laser Light

Have you used a laser today? Chances are you have, but you did not realize it. Lasers are widely used but often hidden from sight. They have hundreds of different uses, from making art to performing open heart surgery.

Lasers are not found in nature. They make a special type of artificial light. Unlike a **light** bulb, the light that comes out of a laser is only one **wavelength** and travels in one direction in a narrow beam. Also, the waves in a laser have their peaks and troughs lined up—they are in phase with one another. This makes laser beams very bright and narrow. Laser beams enable a lot of light **energy** to be focused in a very small area.

Because lasers form a tight beam, the light they make can travel long distances. It is this property of lasers that is most widely used. Laser emitting diodes are used to transmit signals via fiber optics—through thin glass strands. So, if you have watched TV or used a computer or cell phone, you have interacted unknowingly with a laser. If you went to the store, it is likely that a laser scanned your purchase.

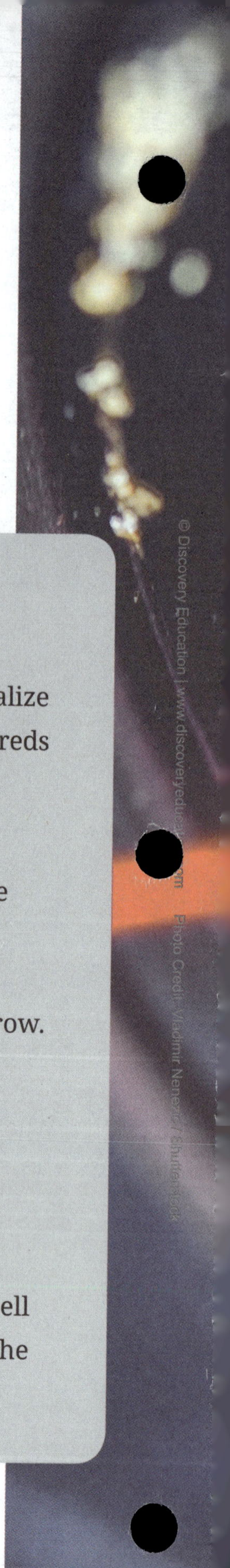

Lasers have a lot of other uses. Their high-energy beams can be used to precisely cut materials. In these cases, extreme heat is produced as the laser light is absorbed by the material. The heat provides the energy needed to make cuts in the materials. This makes lasers very useful in surgery where clean and precise surgical cuts are essential. Scientists use lasers to study rocks on Mars and the atmosphere on Earth. Lasers are also used in satellites to map the surface of planets, bounced off mirrors on the moon, and used by rock music groups to accompany their concerts.

In this concept, you will learn about how mechanical and electromagnetic waves—including laser light—interact with **matter** and how these interactions are used in a wide range of different technologies.

Complete the graphic organizer.

ENGAGE

Activity 4
Evaluate

Quick Code
ca8794s

What Do You Already Know About Waves and Matter?

Waves and Radiation

Can you tell the difference between mechanical waves and electromagnetic waves? Each object demonstrates either a mechanical or electromagnetic wave. **Draw** a line from each object to the appropriate type of wave.

Waves on a Lake

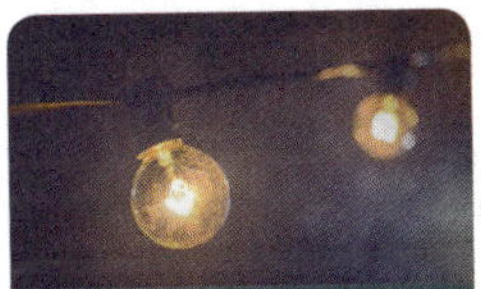
Light Waves from a Light Bulb

Waves from a Drop of Water

Light Waves from the Sun

Light Waves Shining through a Prism

Mechanical Wave

Electromagnetic Wave

Understanding Light

Light behaves differently with different materials.

Draw a line to match each type of behavior with its description.

absorption	light bounces off the material
reflection	light is taken up by the material
transmission	light passes through the material

ENGAGE

Wave Travel in Media

Consider four different types of waves: ocean waves, radio waves, seismic waves, and sound waves.

Draw a line from each phrase to each type of wave that it correctly describes. Some phrases will be used more than once.

transmit energy	Ocean Waves
longitudinal wave	Radio Waves
travel in vacuum	Seismic Waves
mechanical wave	Sound Waves
require a medium	

How Do Mechanical Waves Interact with Matter?

Activity 5

Reason

Quick Code ca8787s

Modeling with Water Waves

In this investigation, you will create a procedure to investigate how water waves will react to different objects in a ripple tank.

What materials do you need? (per group)

- Large bin, with lid
- Glass baking dish (for ripple tank)
- Overhead projector
- Water
- Metric ruler
- Poster board
- Various objects of different shapes that can be placed into water (stones, blocks, rulers, etc.)

Procedure

1. Create a procedure to investigate how water waves react to different objects in a ripple tank.
2. Select your objects and their position in the tank. Observe the waves.
3. Record your observations.
4. Create simple drawings that illustrate your observations.

SEP **Developing and Using Models**

Reflect

What scenarios did you create? How did the waves react?

If you yelled toward a wall, what would the sound waves from your mouth do?

If you shine a flashlight at a tiny hole in a wall, what will happen to the light on the other side of the wall?

EXPLORE

Activity 6

Observe

HELLO!...hello!...

What causes an echo? **Go online** to explore echoes and **answer** the questions that follow.

Quick Code
ca8788s

What causes an echo?

SEP Engaging in Argument from Evidence

Why do echoes only occur with hard surfaces?

What condition produces a softer echo?

EXPLORE

Activity 7
Analyze

Sound Waves and Matter

Quick Code
ca8789s

Read the following text and **watch** the video. Then, **answer** the questions that follow.

Sound Waves and Matter

The process in which waves bounce off a surface is called reflection. When a sound wave in air strikes a surface, it bounces back. The sound is reflected back into the air. Usually sound is only partly reflected, since some of its energy is absorbed by the surface. Depending on the material, some sound may pass through the surface.

When sound is absorbed by a surface, its energy is reduced. The loss of energy reduces the sound wave's intensity, or its loudness. On the other hand, when sound is reflected, it travels back in the direction of the sound source. You might hear the reflection as a reverberation or echo.

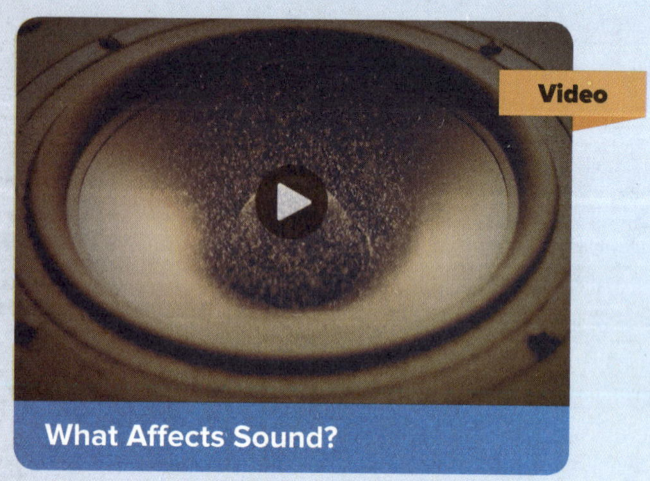

What Affects Sound?

Smooth, hard surfaces generally **reflect** sound waves. Soft, textured surfaces generally absorb sound waves.

When sound waves encounter rough or porous surfaces, they will reflect within the surface. This internal reflection absorbs sound energy. This difference explains why sounds are louder in a tiled bathroom than in a bedroom or other room lined with soft materials such as carpeting. Different materials absorb sound waves with different frequencies.

When sound travels through an object, it causes the object to vibrate. Depending on the medium and its thickness, the sound may travel through it and vibrate the medium on the other side. Consider someone speaking inside a submarine. The voice will vibrate the air, which will in turn vibrate the metal wall of the submarine. The vibrating wall would then cause the water outside the ship to vibrate. These sounds can be detected by listening devices. These effects of sound explain why submarines have a "stealth" mode, during which they run as quietly as possible.

Sound travels at different speeds through different media. In dry air at a temperature of 20°C, sound travels at a speed of about 340 meters per second. You can use the speed of sound in air to calculate the distance of a storm. Lightning and thunder are produced at the same instant, but you see the lightning before you hear the thunder. This difference is because **light** travels almost a million times faster than sound through air. Count the number of seconds after the lightning flash until you hear the thunder crack. Multiply the number of seconds by 340 meters to find how far away the lightning struck.

Sound Waves and Matter *cont'd*

Sound travels much faster through water, at a speed of about 1,480 meters per second. It travels even faster through dense solids, in which the particles of matter are packed very closely together. The speed of sound through steel is about 4,500 meters per second. Temperature also affects how quickly sound travels. All other variables being equal, sound will travel faster when a medium is warmer.

What are the three things that can happen to a sound wave when it strikes an object? **Describe** what each one means.

____

____

Does sound always travel at the same speed? If not, what affects the speed of sound?

____

____

____

____

____

____

____

Activity 8

Observe

Inside the World's Quietest Room

Quick Code
ca8790s

Think about the quietest space you have ever experienced and what materials were used to make the space or to furnish the space. **Watch** the following video and **answer** the question that follows.

Video

Inside the World's Quietest Room

What similarities exist when thinking of your own experience? **Record** your ideas below.

EXPLORE

Activity 9

Evaluate

Sound-Absorbing Materials

Quick Code
ca8910s

You have been promoted at the acoustics company! You have moved from the shipping department to research and development. Your first client presents you with a problem. They moved into a new condo, but the upstairs neighbors make a lot of noise. The client has tried ear plugs, noise-canceling headphones, and white noise, and yet nothing has worked. They can still hear every step their neighbors take. You recommend installing sound-absorbing material in the client's ceiling so that the neighbors' sounds will no longer be heard. **Answer** the following questions.

Describe the properties and structure of the material you recommend.

Photo Credit: MirageC / Moment / Getty Images

Explain in terms of energy and the material's structure how it will physically absorb the sound.

Draw a representation of the material and its geometry to show how sound waves travel after making contact.

EXPLORE

What Happens When Light Is Reflected or Absorbed?

Activity 10

Investigate

Quick Code
ca8791s

Hands-On Investigation: Reflection from a Flat Mirror

In this investigation, you will explore the relationship between the angle at which you shine light on a mirror and the angle at which the light is reflected from the mirror.

Predict

How do you predict the angle that you shine light on a mirror will affect the angle that the light is reflected from the mirror?

SEP **Constructing Explanations and Designing Solutions**

What materials do you need? (per group)

- Laser pointer
- Flat mirror
- Colored pencils
- Piece of white paper
- Protractor
- Ruler

HANDS-ON INVESTIGATION

EXPLORE

Procedure

1. Place the paper on a flat surface.
2. Place the base of the mirror on the paper so that it stands up vertically.
3. Hold the laser pointer, mirror, and white paper so that the light strikes the mirror and reflects onto the paper.
4. Draw the position of the mirror's surface and the path of the beam onto the paper.
5. Use the ruler and protractor to draw the line of the light to and from the mirror.
6. Draw a line perpendicular to the mirror at the point where the beam hits the mirror.
7. Determine the angles relative to the perpendicular at which the light strikes and is reflected off the mirror in a table like the one shown.
8. Repeat with different angles.

Test Number	Angle at Which Light Strikes the Mirror (degrees)	Angle at Which Light is Reflected off the Mirror (degrees)
1		
2		
3		
4		
5		

Reflect

How did the angle at which the light strikes the mirror compare to the angle at which the light is reflected from the mirror in each test? How does this compare with your prediction?

Based on the results of the experiment, create a "rule of reflection" that explains the relationship between these angles.

What problems did you encounter when trying to measure the angles of the beam of light?

How could you make modifications to this experiment to reduce any problems that you encountered?

EXPLORE

Activity 11

Observe

The Law of Reflection

Quick Code
ca8792s

Watch the following video and **answer** the questions that follow.

The Law of Reflection

SEP Developing and Using Models

What is the law of reflection defined at the end of the video?

How does your self-generated law of reflection compare to the definition presented in the video?

EXPLORE

Activity 12

Observe

Quick Code
ca8793s

Reflection from Other Objects

Look at the object your teacher gives you.

Draw a simple diagram that explains how you can see this object. Then, **watch** the video and **answer** the questions that follow.

Reflection and Absorption of Wavelengths

SEP Developing and Using Models

CCC Cause and Effect

When we see colors, what are our eyes detecting?

What colors produce white light?

When you look at a red object in white sunlight, why does it look red?

When you look at a black object in white sunlight, why does it look black?

Complete the sentence: The color you see when you look at an object depends on...

EXPLORE

Activity 13

Investigate

Quick Code
ca8911s

Hands-On Investigation: Light and Temperature

In this investigation, you will measure the temperature of water inside of vessels of different colors when they are hit with white light. You will use the data to determine how color can affect an object's temperature when hit with white light.

Predict

How will color affect an object's temperature when hit with white light?

SEP Developing and Using Models

SEP Engaging in Argument from Evidence

CCC Cause and Effect

What materials do you need? (per group)

- Jars, glass, 4 oz
- Flashlights
- Thermometers
- Graduated cylinder, 250 x 2 mL
- Graph paper
- Colored pencils
- Paint (for jars)
- Batteries, size D

HANDS-ON INVESTIGATION

EXPLORE

Procedure

Design your experiment, including which variables will be controlled, which variables will be independent, and what the dependent variable is.

Carry out your experiment and record your data in a data table similar to the one shown.

Temperature at:	5 minutes	10 minutes	15 minutes	20 minutes	25 minutes
White					
Black					
Red					
Green					
Blue					

Use colored pencils representing the color of the vessels to construct a line graph. Show the temperature of the water on the *y*-axis and the heating time on the *x*-axis.

Reflect

Explain what was happening to the white light when it hit each of your five vessels. You may use diagrams if you would like.

From your investigation, which color reflected the most light and which color absorbed the most light? Explain your reasoning.

EXPLORE

What Happens When Light Passes through Matter?

Activity 14

Observe

Transparent, Translucent, and Opaque

Quick Code
ca8796s

Watch the video and use what you learn to **complete** the activity that follows.

Transparent, Translucent, and Opaque

Create Frayer models for each term: *transparent, translucent,* and *opaque.*

EXPLORE

What happens to a ray of light as it is reflected off a mirror and into a glass of water?

Activity 15

Investigate

Hands-On Investigation: Transparent, Translucent, and Opaque Materials

Quick Code
ca8912s

EXPLORE

In this activity, you will use a flashlight to observe the opacity and reflectivity of various objects.

Predict

Which objects do you think will be opaque or reflective?

CCC Structure and Function

What materials do you need? (per group)

- Flashlight
- Batteries, size D
- Wax paper
- Plastic wrap
- Notebook paper
- Tissues
- Construction paper
- Stapler
- Spoon, metal teaspoon

HANDS-ON INVESTIGATION

Procedure

Create a plan for determining the opacity and reflectivity of objects. **Describe** the procedure you will follow.

Carry out your procedure and **record** your data in the chart.

Material	Prediction: Transparent, Translucent, or Opaque	Observations: Transmitted, Reflected, or Absorbed	Actual Classification: Transparent, Translucent, or Opaque
Wax Paper			
Plastic Wrap			
Notebook Paper			

EXPLORE

Reflect

How did your predictions compare to your observations? Were you surprised by any of the results?

What are the similarities between objects in each category: transparent, translucent, and opaque?

How do your observations support the theory that light is a wave?

Activity 16

Analyze

Translucent and Transparent Materials

Quick Code
ca8913s

Read the following text. **Highlight** any statements you observed as evidence during the Transparent, Translucent, and Opaque Materials activity.

EXPLORE

Translucent and Transparent Materials

All surfaces and materials reflect some light but only opaque objects absorb the rest of the light that strikes them. Some objects allow light to enter and pass through them. These objects are either translucent or transparent.

Light can travel through some materials, such as glass, **air**, and water. We say that light is transmitted through these substances. When light strikes a glass window, some of the light is transmitted through the glass, while some of it is reflected back. This is the reason why you can see an object at the bottom of the pool, but you can also see your reflection at the surface of the water. The amount of light that is transmitted or reflected depends on the properties of the new medium. It also depends on the frequency, or color, of the light. Some materials will absorb some wavelengths of light but allow others to pass through. These materials are called filters. Color filters are widely used in devices and in photography. A red filter makes everything look red because it absorbs all wavelengths but red.

Activity 17

Observe

Refraction

Quick Code
ca8799s

Watch the video and **answer** the questions that follow.

What do you notice about the ray that bounces off the block?

What happens to the ray when it enters the block?

What happens to the ray when it leaves the block?

CCC **Cause and Effect**

Draw your observations in the space provided.

Refraction is the bending of light (or other waves) that occurs when light moves from one transparent medium to another.

Draw a glass block. Then, watch the video again. As you watch, **draw** your observations of the following components shown in the video and then **answer** the question that follows.

- incident ray
- reflected ray
- refracted ray in the block
- refracted ray outside the block

Why does refraction occur?

EXPLORE

Activity 18

Analyze

How Waves Travel around Objects

Quick Code
ca8800s

Read the text. As you read, **circle** the key words from the lesson.

Then, **write** a sentence that explains how the terms you circled are related to the learning objectives.

How Waves Travel around Objects

Imagine you are standing near the corner of a building in the school yard. Around the corner, out of direct sight, you can hear a student playing rap music on a phone. Do you think the **sound** from the phone is passing through the two walls of the building, or can you hear it around the corner? Probably some of both, but most of the sound is getting around the corner by a process called **diffraction**.

When a wave hits an obstacle and the wave bends around the obstacle, it is called diffraction. Diffraction is most noticeable when the **wavelength** of the wave is close to the size of the obstacle.

All waves are bent when they pass through a slit, causing them to spread out as waves like ripples on a pond. When a wave passes through two slits, the wave bends around each slit. This results in two waves, one from each slit. A diffraction pattern is formed when the two waves sum additively to form peaks and troughs.

CCC Patterns

In the case of **light** waves, this can be observed as areas of maximum intensity and minimum intensity on a screen. The maximum intensity areas are places where the crests of the two waves meet to produce constructive interference. The new wave has a higher overall intensity. The minimum intensity areas are places where the peaks of one wave meet the troughs of the other wave to produce destructive interference. The new wave has a lower overall intensity.

EXPLORE

Now, **write** your sentence.

Activity 19

Evaluate

Waves and Matter

Quick Code
ca8802s

Waves produce different phenomena depending on how they interact with matter. **Write** the appropriate term to identify each specific phenomenon illustrated in the diagrams.

Illustration of Wave Phenomenon	Term
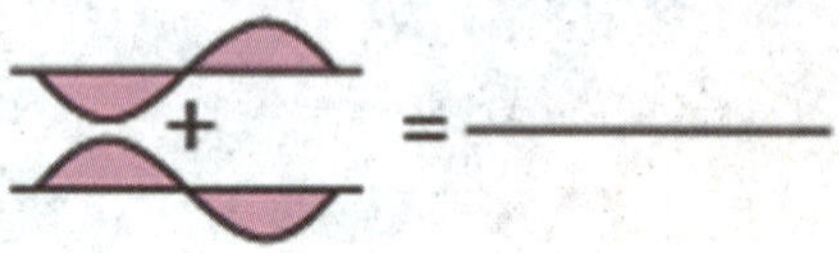	

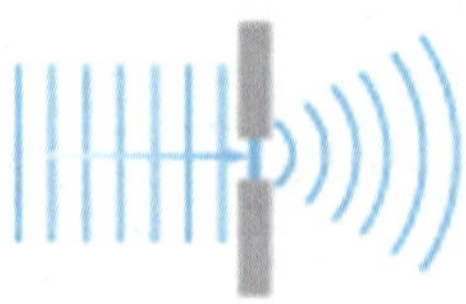	
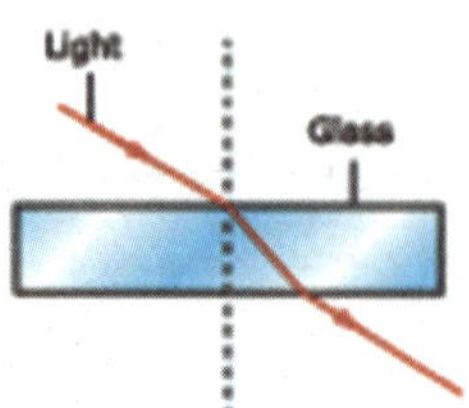	

Activity 20

Solve Problems

STEM

Quick Code
ca8803s

Periscope

Periscopes are often associated with military use. For example, sailors use periscopes to see above the water while they are in a submarine below the ocean surface. However, periscopes can be used in many other situations as well. Spectators at golf tournaments can use periscopes to see over the crowd onto the green. Periscopes can be used by home inspectors to look into spaces that are difficult to see. If an object is blocking your line of sight or if you need to see something that is out of your line of sight, a periscope may be just what you need.

In the activities that follow, you will **research, design,** and **construct** a functional periscope.

Periscope

SEP Constructing Explanations and Designing Solutions

EXPLORE

Mirror, Mirror

Periscopes are used to see over objects, so the end you look into is lower than the end you see out of. Mirrors are placed inside the periscope to direct the light from the object to your eye. Where should mirrors be placed in order to direct the light reflecting off the object down to the eye? **Circle** all the areas that apply.

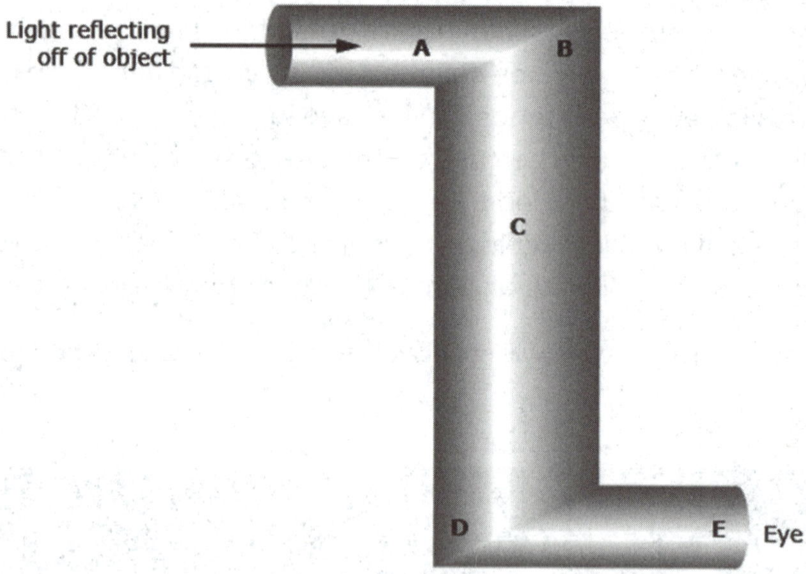

Criteria for Success

What criteria will your periscope need to meet to be considered a successful periscope? **Include** what it will need to accomplish and which pieces of the periscope are critical. **Describe** how you will test your periscope to determine that it works correctly.

EXPLORE

Detailed Design

The angles of the mirrors in your periscope will affect how the light travels. **Create** a geometrically accurate sketch of your periscope. On your sketch, **show** the path the light would follow as it travels through your periscope. Be sure to **use** a protractor to **measure** and **label** the angles at which light strikes each mirror. **Use** a straightedge to **draw** the path of the light beam. You may need to redraw the angle of your mirrors a few times so they will reflect light along the correct path.

Write a short paragraph to describe your model, including the degrees of the angles.

Construct, Test, and Compare

Build your periscope according to your design and **test** it according to the test you chose.

How did your design differ from those of your classmates?

Were there any types of design that did especially well or especially poorly?

If you were to make a second periscope, what would you change about your design?

EXPLORE

Activity 21

Record Evidence

Laser Art

Quick Code
ca8804s

As you worked through this lesson, you investigated and gathered evidence about waves and matter. Now, take another look at the video Laser Art, which you first saw in Engage.

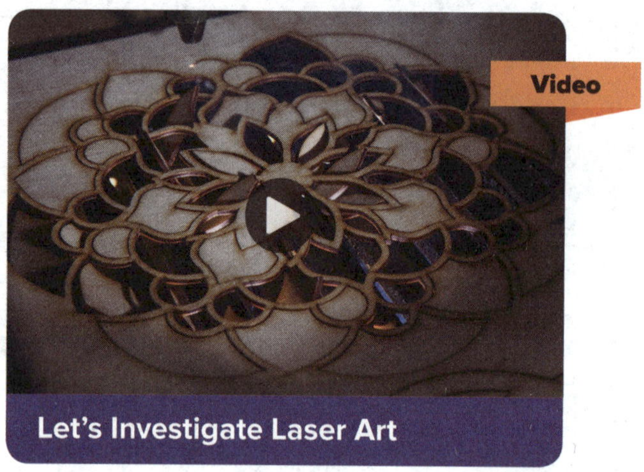

How has your understanding of the Laser Art video changed?

Read the Can You Explain? question from the beginning of this lesson.

Can You Explain?

What happens to a ray of light as it is reflected off a mirror and into a glass of water?

Use your new understanding of the Laser Art video to write a scientific explanation answering a question.

1. **Choose** a question. You may choose to answer the Can You Explain? question or one of the questions you wrote at the start of this lesson.
2. Use the following information to help you write your explanation.

Construct an explanation that uses one or more models or representations as evidence. Models can include graphs, charts, drawings, photographs, or other types of graphics. Prior to writing your scientific explanation, think about what sort of models would best serve as evidence to support your claim. You can either draw your representations or print them out and attach them to the explanation.

EXPLAIN

SEP **Constructing Explanations and Designing Solutions**

What happens to a ray of light as it is reflected off a mirror and into a glass of water?

EXPLAIN

STEM in Action

Activity 22

Analyze

X-Ray Crystallography

Quick Code
ca8805s

Read the passage and **look** for details that show how X-ray crystallography works.

X-Ray Crystallography

You are probably aware that X-rays can be used in the medical field for imaging teeth and bones. However, did you know that X-rays can also be used in a wide variety of scientific fields? X-rays can be used to determine how the atoms of materials with very ordered structures are arranged. These materials are called crystalline materials, and the process is called X-ray crystallography. X-ray crystallography has applications for biology, geology, and materials science. Many scientists become experts in this technique and use it as a large part of their career. What are some careers that involve crystallography?

Video

The Double Helix

When you think of crystals, you probably think of minerals. It might seem impossible, but X-ray crystallography was actually used to determine the structure of DNA, the molecule that makes us who we are. How was the same scientific technique used to study mineral crystals used to solve the mystery of what DNA looks like?

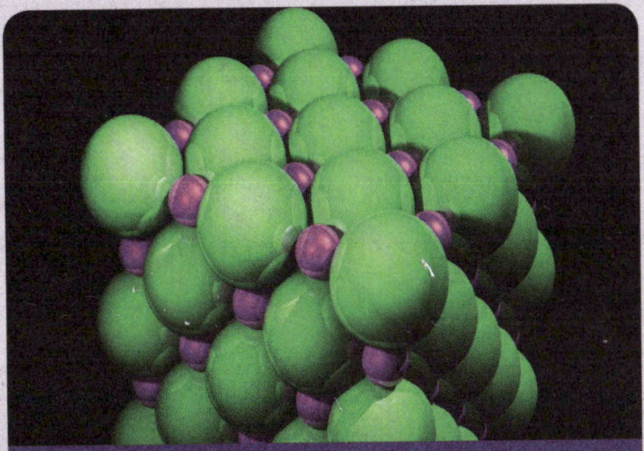

Sodium Chloride Crystal and Crystal Lattice Structure

Crystallography is the study of the arrangement of particles and bonding of atoms in crystalline solids. Compounds formed by ionic bonds form crystal lattices in which the arrangement of the ions is repeated in a regular pattern. The ionic compounds do not contain individual molecules.

Crystals have characteristic geometrical shapes that are related to the arrangements of ions. In early studies, crystallographers learned about the shapes of these compounds by measuring the angles between the faces of crystals.

ELABORATE

X-Ray Crystallography *cont'd*

They used the optic properties of crystals, measuring the diffraction angles and frequencies of light that passed through them. Modern X-ray diffraction crystallography, however, changed how scientists study crystals.

When an X-ray beam is directed at a crystal, a diffraction pattern is produced. The diffraction pattern is related to the arrangement of the particles in the crystal. Using the diffraction pattern to predict the arrangement of a compound's ions can be tricky. The task has been compared to building a piano based only on the sounds it makes while it is falling down a flight of stairs. It is a complex procedure involving mathematical skills. This once tedious task, however, has been made much easier by advances in computer technology.

Although ionic crystals seem like the obvious subject of study in this field, the technique can even be applied to studies of large, complex organic molecules. In fact, the double helix structure of DNA was deduced, in part, from crystallographic data. Today, crystallographers work in many fields, including chemistry, physics, geology, biology, materials science, medicine, and metallurgy.

A person choosing a career as a crystallographer can expect to work in a laboratory setting. The work involves sophisticated technical instruments and computers. Chemistry, physics, and biology are important high school courses for a person interested in a career in crystallography. Mathematics up to and including calculus is also necessary. Undergraduate college preparation includes chemistry, physics, mathematics, and advanced computer science and programming. Laboratory experience will be very valuable.

Other courses needed might include biology, geology, or materials science, depending on the field in which the student hopes to work. Many, but not all, crystallographers acquire an advanced degree, especially those who plan to work as research scientists. At the graduate level, the student can focus on a specialty and refine the skills related to that area of the field. Some employment opportunities include research, scientific journal editing, materials engineering, or sales and marketing.

ELABORATE

X-Ray Crystallography

Number the steps of X-ray crystallography shown here in order to create a diagram showing how X-ray crystallography works.

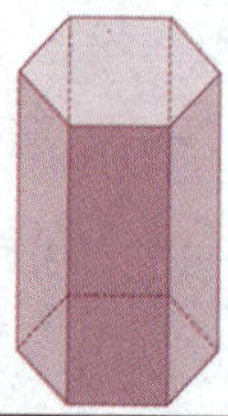

Step 1

Step 2

Step 3

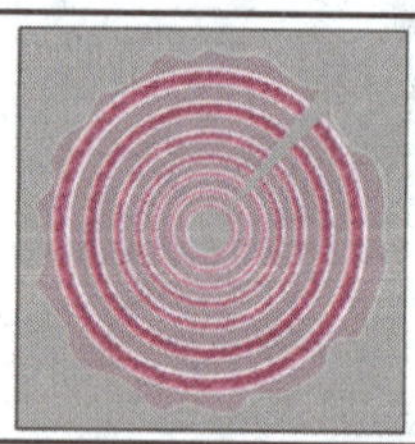

Step 4

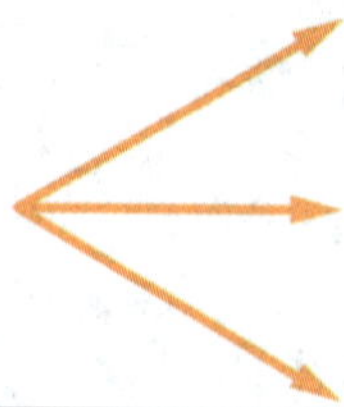

Step 5

ELABORATE

Activity 23

Concept Review

Review: Waves and Matter

Quick Code
ca8914s

Now that you have completed the objectives for this concept, **review** the core ideas you have learned. Record some of the core ideas below.

Core Ideas

Talk with a Group

Now, think about the video Sensing Biodiversity from the Sky, which you saw in Get Started. Discuss how what you have learned about waves and matter can help you understand how the airborne observatory monitors life below.

EVALUATE

CONCEPT 4.3

Warming Earth

Student Objectives

By the end of this lesson:

- [ ] I can analyze and interpret Earth's energy budget to describe how energy is conserved and to predict the effect of change in one system on other systems.
- [ ] I can develop a model that describes how Earth's shape and rotation result in patterns of distribution of solar radiation.
- [ ] I can construct an explanation of the many factors that cause seasonal change on Earth and why those seasons vary over Earth's surface.
- [ ] I can argue from evidence that factors such as the shape and tilt of Earth impact climate, biodiversity, and organism distribution.
- [ ] I can obtain, evaluate, and communicate how changes to Earth's climate cause change in biodiversity and organism distribution.

Key Vocabulary

atmosphere, biodiversity, chlorophyll, deciduous forest, energy, fusion, hydrogen, infrared, matter, photosynthesis, solar energy, sun, topography, tundra

Quick Code
ca8806s

Activity 1

Can You Explain?

How does the shape of our planet and its axial tilt impact the distribution of solar energy and life upon its surface?

Quick Code
ca8807s

Activity 2

Ask Questions

The Sun

Quick Code
ca8808s

Watch the following video, and then **answer** the questions that follow.

Video

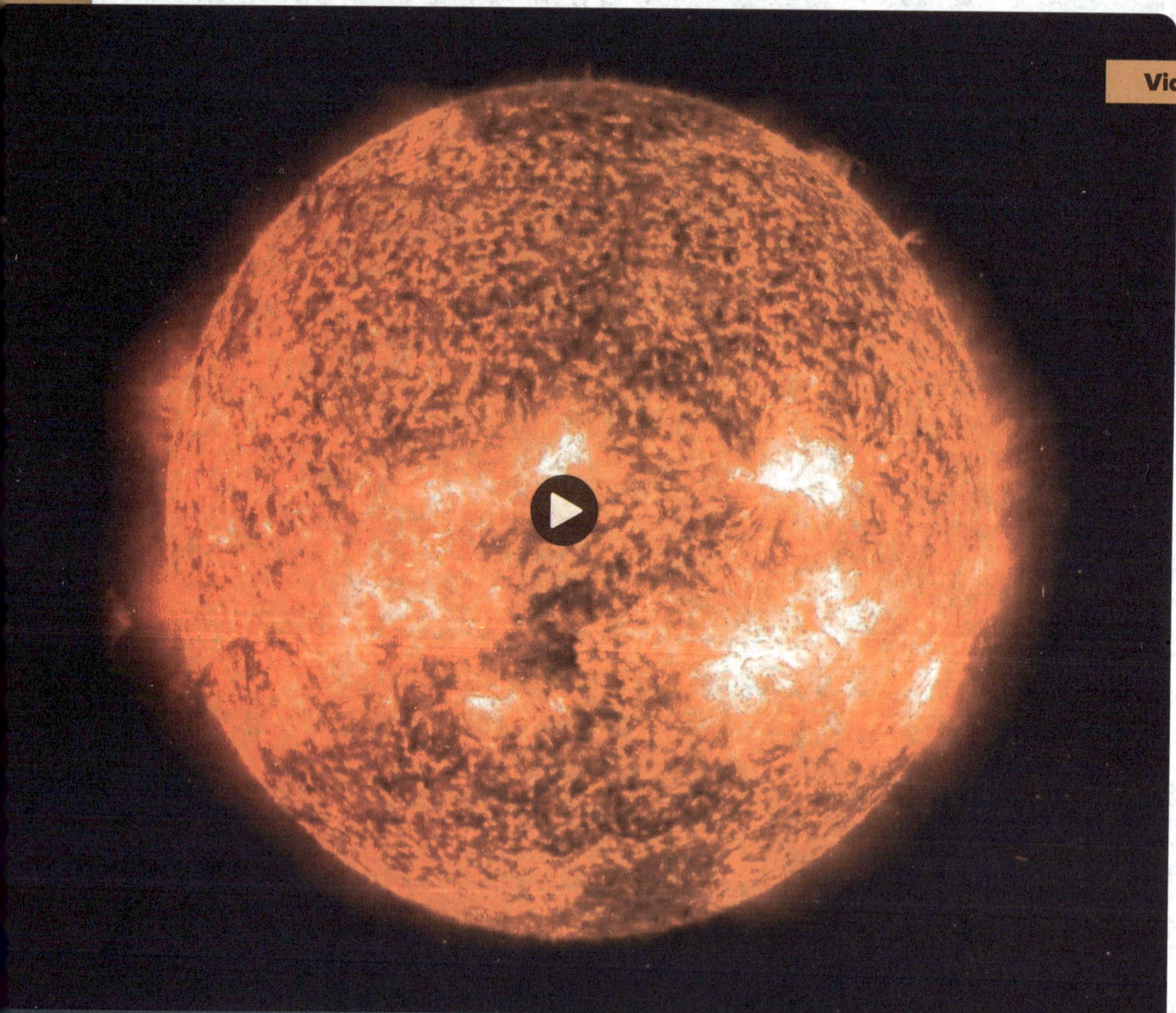

Let's Investigate the Sun

What causes the sun's energy?

How might the sun's energy affect Earth?

What questions do you have about the sun's energy? Write them below.

ENGAGE

Activity 3

Analyze

The Violent Sun

Quick Code
ca8809s

Read the following text. **Highlight** the text in different colors to identify:

- where the sun's energy is produced
- the process that produces this energy
- the type(s) of energy released by the sun

Then, **watch** the video to learn more about how the sun makes energy.

What do you think happens to solar energy when it reaches Earth?

____

____

____

Identify and **list** the importance of the sun's energy in natural and human-made systems.

____

____

____

____

____

____

____

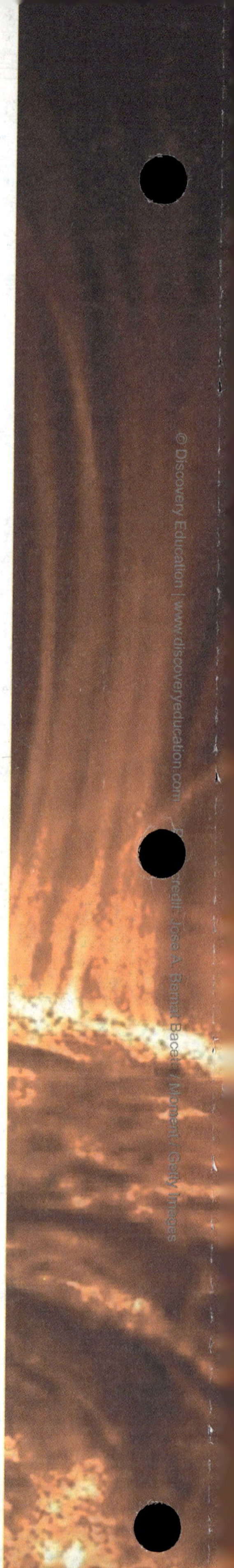

ENGAGE

The Violent Sun

The **sun** is the main source of **energy** for Earth. The sun's energy comes from its core. Within the sun's core, hydrogen is converted into helium in a process called nuclear fusion. Fusion releases a tremendous amount of energy. Because of the sun's massive gravitational pull, most of the energy cycles back into the sun, but some of the energy moves to the sun's surface, where it is released into space.

The sun emits energy as electromagnetic radiation. All electromagnetic radiation travels as waves, and waves carry energy. Electromagnetic waves consist of an electric field and a magnetic field. Each wave travels through empty space in a straight line, or ray. Visible light is a form of electromagnetic radiation. Light rays, like all electromagnetic waves, can travel indefinitely through space. This is why we can see distant stars that are trillions of miles away.

Only a small amount of the sun's energy reaches Earth. Yet without this energy, life could not exist on our planet.

In this concept, you will learn what happens to the energy the sun generates, including the various types of energy Earth receives from the sun. You will also explore how this distribution of energy over Earth's surface determines the climate and distribution of life on the planet.

Video

Sun: Energy Machine

Activity 4

Evaluate

Quick Code
ca8810s

What Do You Already Know About Warming Earth?

Light Energy

Which of these is the same form of energy as light? **Select** the correct answer.

- ☐ Radio waves
- ☐ Sound waves
- ☐ Electricity
- ☐ Potential energy

Comparing Different Types of Electromagnetic Waves

The sun emits three main types of electromagnetic radiation: visible light, infrared radiation, and ultraviolet radiation. **Consider** how the properties of each of these types of radiation are similar and different, and then **write** the properties in the appropriate columns in the chart.

Properties of Electromagnetic Radiation

- frequency range
- wavelength range
- visible to humans
- speed
- energy
- types of interaction with matter

Same for All Types of Waves	Different for Each Type of Wave

ENGAGE

Next, **respond** to the following question.

Explain what is meant when we say that the energy (for example frequency, wavelength) is different for each type of wave.

Seasonal Differences

Classify each characteristic as "higher" or "lower" during the summer versus the winter. **Write** your response in the appropriate space below.

	Summer	Winter
Height of the Sun at Noon		
Angle of Sun's Rays		
Length of Shadows at Noon		
Time the Sun Is in the Sky		

ENGAGE

What Happens to Electromagnetic Radiation from the Sun When It Strikes Earth?

Activity 5

Observe

Quick Code
ca8811s

Earth's Energy Budget

Answer the following question.

What happens to sunlight when it reaches Earth?

SEP Developing and Using Models

Examine the following image.

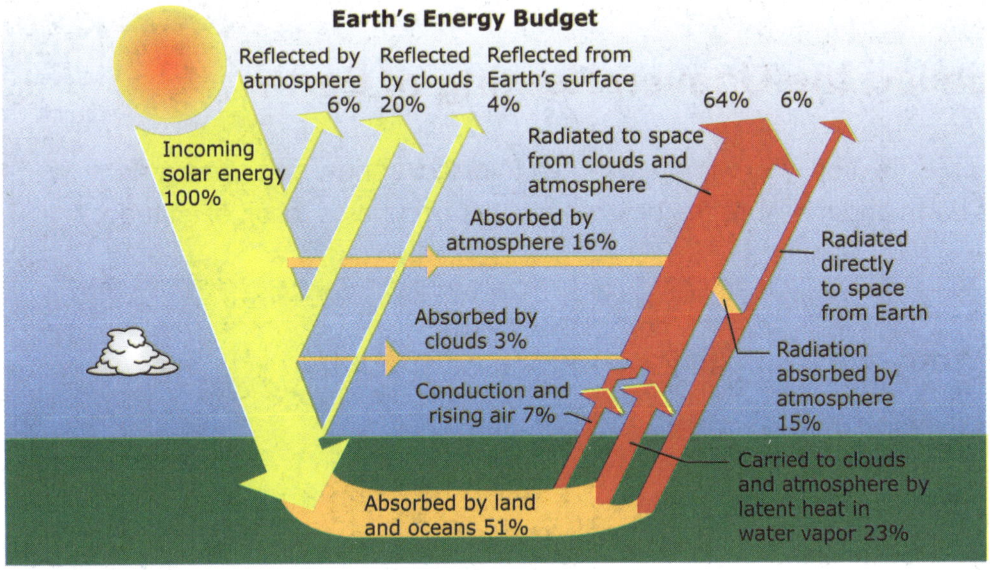

EXPLORE

Determine what proportion of the sun's energy ends up being absorbed by Earth and what proportion ends up being released back into space.

What would happen to Earth if all the energy obtained from the sun did not equal energy released into space?

Activity 6

Reason

Modeling the Uneven Heating of Earth

Quick Code
ca8812s

In this investigation, you will use a model to discover how solar energy strikes Earth and measure the differences that occur as a result of latitude.

What materials do you need? (per group)

- World map or globe
- Chalkboard or whiteboard
- Basketball
- Bowls, plastic
- Chalk
- Flashlight
- Batteries, size D
- Measuring tape
- Metric ruler
- Cardboard paper towel tube
- Scissors
- Graph paper

Procedure

1. Use chalk to mark lines on the basketball that represent the equator, mid, and high latitudes.
2. Shine light directly on each of the marked latitudes and compare the resulting areas of light.
3. Write a claim. What do you predict will be the result of the investigation?

____

____

SEP **Analyzing and Interpreting Data**

4. Shine the flashlight from a fixed position onto the equator.

5. Outline the direct light from the flashlight.

6. Repeat the procedure when the flashlight is aimed at the other latitudes.

7. Record the size of the ellipse created at each step.

8. Calculate the area of each ellipse using the formula (Area = $a \times b \times \pi$).

9. Plot the results on graph paper. What data did you record at this step?

____

Develop a conclusion for your investigation.

____

____

____

____

EXPLORE

Reflect

What pattern do you see in your graph?

How does solar radiation (energy per unit area) change with latitude?

How could you use your model to predict the amount of solar radiation experienced at a latitude of 15 centimeters above the equator on your model?

Activity 7

Analyze

Uneven Distribution of Solar Energy

Quick Code
ca8813s

Read the text and **watch** the videos. Then, **answer** the following question.

Although its incoming and outgoing radiation is balanced, Earth does not heat evenly. In fact, some parts of Earth, especially those near the poles, lose heat faster than they gain energy from the sun and, for part of the time, some parts receive no energy at all.

What are the factors that affect the amount of solar energy that reaches different parts of Earth?

EXPLORE

Uneven Distribution of Solar Energy

Day and Night

Temperature on Earth

There are several reasons for this. First, only one half of Earth is bathed in solar radiation and experiences daytime. The other half of Earth is in darkness. Because Earth rotates on its axis, most parts of Earth experience day and night once every 24 hours.

Because Earth is spherical, the amount of solar energy that reaches a given area of its surface decreases as you move away from where the sun is directly overhead at midday. This means that as you move toward the poles, there is less solar energy falling on Earth per unit area than at lower latitudes. This is because the same amount of solar energy is spread out over a larger area. The amount of solar energy a region receives (per unit area) affects its temperature. This explains why polar regions have much colder climates than equatorial regions. The amount of incident solar radiation is only one factor that impacts temperature on Earth. Other factors, such as proximity to the ocean, the nature of the landscape, and cloud cover, are also important.

What Causes Seasons?

Activity 8

Observe

To Every Season

Go online to explore the reason for seasons. **See** how conditions change as Earth revolves around the sun by choosing a location and time of year to observe.

Answer the questions that follow.

Quick Code
ca8815s

EXPLORE

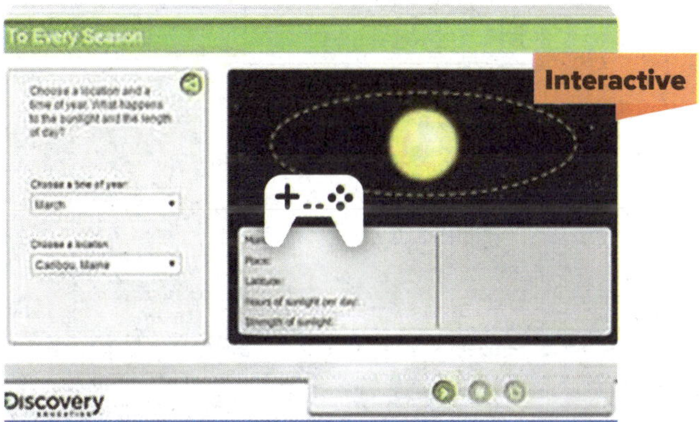

To Every Season

How to Use This Interactive

1. **Read** the Introduction and **click** the Continue button.
2. **Select** a location and a time of the year. **Click** on the Play button.

SEP Developing and Using Models

SEP Analyzing and Interpreting Data

CCC Cause and Effect

3. **Watch** as Earth revolves around the sun to the chosen time of the year. **Read** the data, recording the hours of sunlight per day and strength of the sunlight in the Data Chart. **Note** the tilt of Earth.
4. **Continue** with all combinations.

Observations

How was Earth tilted when each location had the strongest sunlight?

Why was the strength of sunlight generally greater for Miami?

Why does Caribou have more hours of sunlight in June than Miami?

Seasons		Caribou, Maine	Miami, Florida
March	Hrs. of sunlight		
	Strength of sunlight		
June	Hrs. of sunlight		
	Strength of sunlight		
September	Hrs. of sunlight		
	Strength of sunlight		
December	Hrs. of sunlight		
	Strength of sunlight		

Now, **watch** the video to review the causes of the seasons.

The Change of Seasons

What causes the seasons to change?

EXPLORE

Activity 9

Analyze

How and Why Seasons Differ

Quick Code
ca8818s

Read the following text and **highlight** evidence that supports your response to the following questions:

- How is it that different seasons can exist at the same time on Earth?
- How do seasons differ in the Northern and Southern Hemispheres?
- How does latitude affect the nature of the seasons?

How and Why Seasons Differ

Seasons are not the same everywhere on Earth at the same time. In fact, the seasons in the Northern and Southern Hemispheres are always opposite each other. In the Northern Hemisphere, the summer solstice occurs in June and the winter solstice in December. In the Southern Hemisphere, the dates are reversed. Whether the equinox is vernal or autumnal also depends on which hemisphere you are in.

In July in the United States, the weather is warm, and the sun's angle means it sets late in the evening. The same is true in Australia in December, making Christmas a traditional beach holiday for many families there. Both the United States and Australia experience warm summers and cool winters—as well as spring and fall—but at opposite times of year. However, not all places on Earth experience four distinct seasons with different weather patterns.

EXPLORE

Where you are on Earth in terms of latitude also makes a difference in what kinds of seasons you experience. In some places, seasonal changes are extreme, while in others they are not very pronounced. How extreme seasonal differences are is partly a function of latitude. The farther north or south you are, the more variation you will experience in hours and intensity of sunlight.

Generally, the closer a region is to Earth's equator, the less difference there is between seasons. The equator always receives relatively steady sunlight, no matter which way Earth is tilted. At the equator, the length of the day and the angle of the sun at noon—the height the sun reaches in the sky—do not change much. This is one reason why in North America, southern states tend to have milder winters. Of course, other climate factors can still create dramatic seasonal differences, such as rainy or dry seasons.

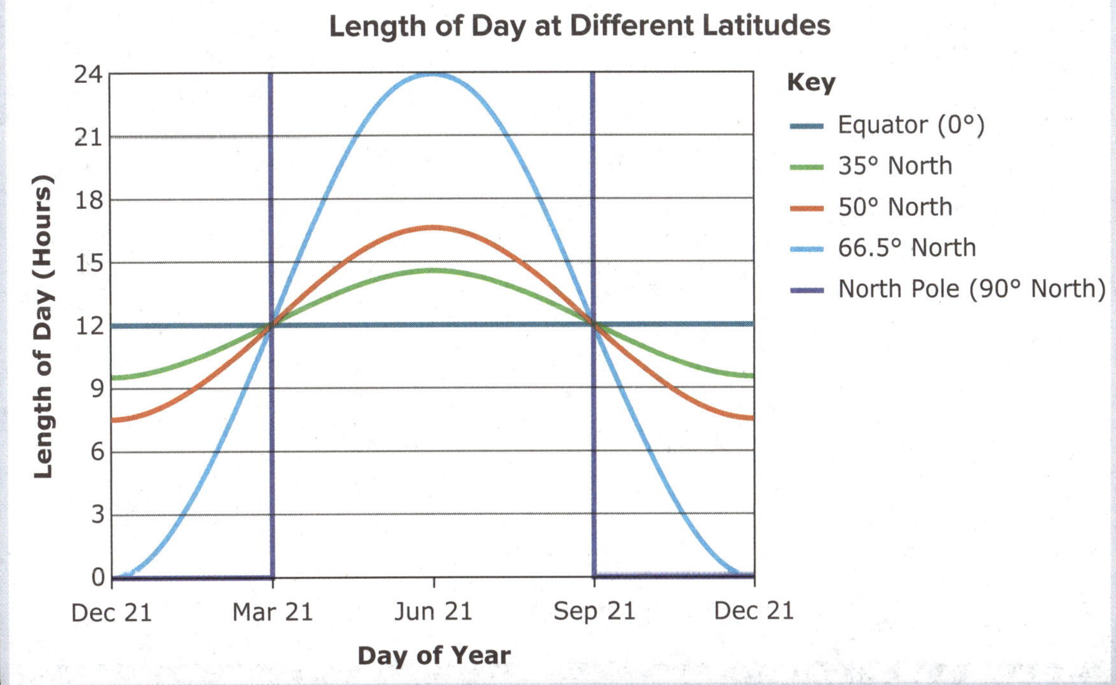

How is it that different seasons can exist at the same time on Earth?

How do seasons differ in the Northern and Southern Hemispheres?

How does latitude affect the nature of the seasons?

Activity 10

Evaluate

Seasons and Latitude

Quick Code
ca8817s

Examine the diagram, and then **answer** the questions.

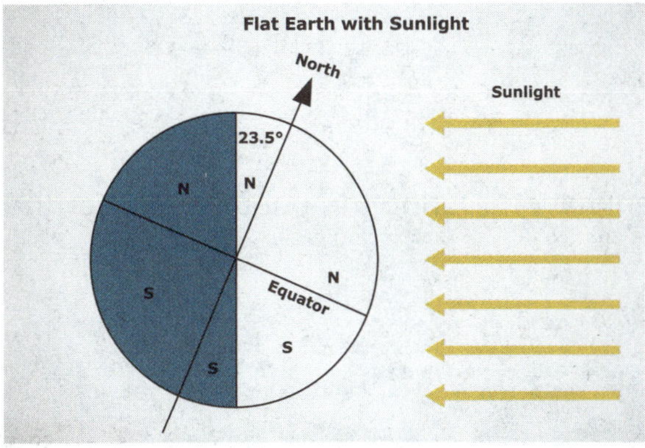

In the diagram, what season is it in the Southern Hemisphere? How do you know?

In the diagram, what location on Earth is receiving the most hours of sunlight? Why?

EXPLORE

How Does the Distribution of Solar Radiation Influence Climate, Biodiversity, and the Distribution of Organisms?

Activity 11

Observe

Biomes

Quick Code
ca8820s

Watch the following animation. During the pauses, **answer** the following questions.

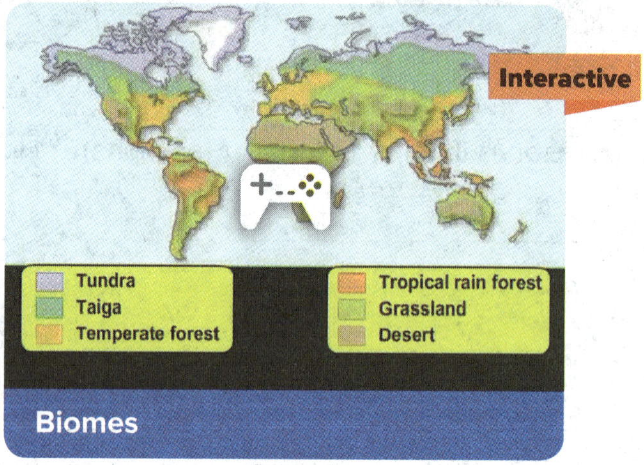

Why are different species found in each of these places?

How do species differ as we move from the poles to the equator?

What role does the distribution of solar radiation play in determining the distribution of organisms?

EXPLORE

Activity 12

Solve Problems

STEM

Quick Code ca8821s

What If Earth Tilted More?

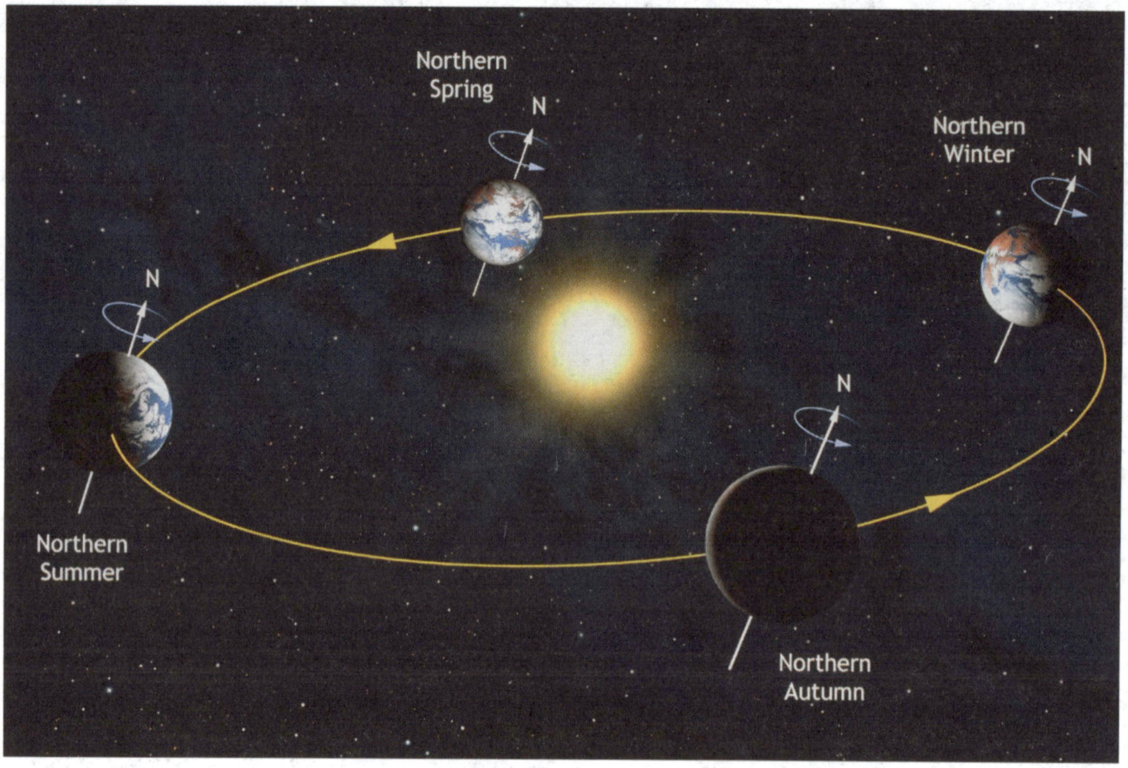

Suppose the angle of Earth's tilt increased. **Think** about the effects this increase might have on Earth's weather, and then **answer** the following questions.

Tilt of Earth

An increase in the angle of Earth's tilt would likely have which of these effects? **Select** all that apply.

- ☐ During winter, the sun would set at an earlier time.
- ☐ During summer, the sun would set at an earlier time.
- ☐ During winter, average temperatures would increase.
- ☐ During summer, average temperatures would increase.
- ☐ Earth would take more time to complete one rotation on its axis.
- ☐ Earth would take more time to complete one revolution around the sun.

Increasing the Tilt

Describe one problem for humans that likely would result from an increase in the angle of Earth's tilt. **Explain** how engineers might solve this problem.

EXPLORE

Activity 13

Record Evidence

The Sun

Quick Code ca8825s

As you worked through this lesson, you investigated and gathered evidence about solar radiation and Earth's surface. Now, take another look at the video The Sun, which you first saw in Engage.

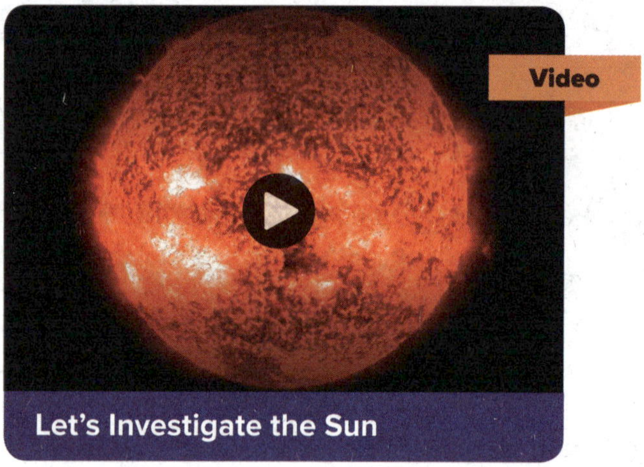

How has your understanding of the sun changed?

Read the Can You Explain? question from the beginning of this lesson.

Can You Explain?

How does the shape of our planet and its axial tilt impact the distribution of solar energy and life upon its surface?

Use your new understanding of the sun to write a scientific explanation answering a question.

1. **Choose** a question. You may choose to answer the Can You Explain? question or one of the questions you wrote at the start of this lesson.
2. Use the following information to help you write your explanation.

Construct an explanation that uses one or more models or representations as evidence. Models can include graphs, charts, drawings, photographs, or other types of graphics. Prior to writing your scientific explanation, think about what sort of models would best serve as evidence to support your claim. You can either draw your representations or print them out and attach them to the explanation.

SEP **Engaging in Argument from Evidence**

EXPLAIN

EXPLAIN

STEM in Action

Activity 14

Analyze

Engineering Solar Power: Improving Lives

Quick Code
ca8823s

Read the text and **watch** the videos about how solar energy can be used to improve people's lives. While reading, **consider** ways that solar energy could be applied to improve the quality of life in your community.

Engineering Solar Power: Improving Lives

Video

Egypt: Solar Power

Solar energy is abundant, free, and non-polluting. With the development of the solar cell and later the solar panel, humans have found ways to capture the sun's radiant energy and convert it into electricity. When solar cells and panels were invented, they were very expensive, and not many people could afford them. In addition, the idea of using solar energy as a source of electricity was so unfamiliar that many people, companies, and governments did not embrace the idea.

SEP Obtaining, Evaluating, and Communicating Information

However, people have begun to realize clean energy is our future, and the use of solar power has increased tremendously. At the same time, the costs of using solar energy have gone down. Lower costs have made solar power accessible to communities around the world, many of which previously struggled with access to any electricity.

Job opportunities have expanded along with the availability of solar power. For example, small farms that previously did not have access to electricity now have a reliable source of power for important farming tasks, such as irrigation. Farmers now can do more work and make more income. The quality of their lives—and the lives of others—is improved.

In many parts of the world, electricity is produced using fuels such as coal. Electricity produced in this way is sent along long-distance wires to reach homes and businesses.

With solar power, a person, village, or company can make its own electricity simply by buying and setting up solar panels. Once the solar panels are in place, electricity is freely available every day for the life of the panels.

Can you think of other types of work or other community endeavors for which free access to solar power would make a big difference?

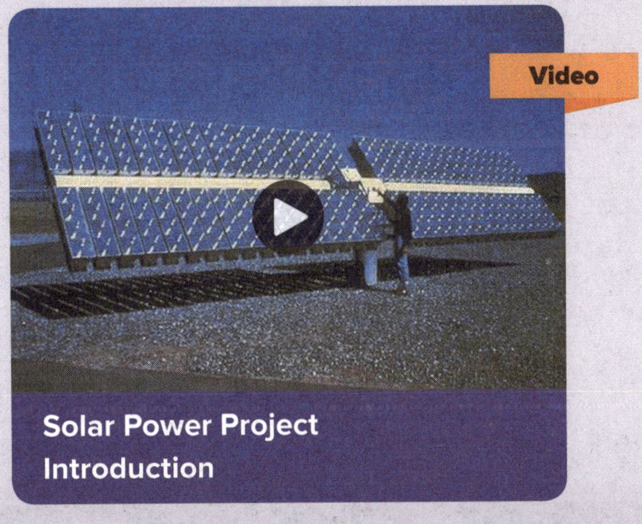

Solar Power Project Introduction

ELABORATE

Swapping to Solar Energy

Now, **think** of ways that solar energy may impact your community. **Answer** the questions individually, then **share** with a partner. Your answers should be specific and well thought out.

How could solar power be useful to individuals or communities that you know? (**Think** about objects or situations that could use solar power but do not.) **Explain** your answer.

How would the use of solar power in the situation you described improve the individual's or community's life or work?

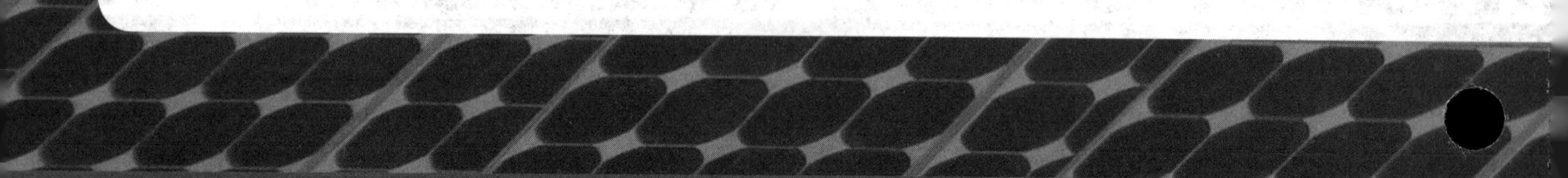

How could a scientist measure how much impact the solar power had on the individual or community? Why might this data be useful to collect and analyze?

Activity 15

Concept Review

Review: Warming Earth

Quick Code
ca8824s

Now that you have completed the objectives for this concept, **review** the core ideas you have learned. Record some of the core ideas below.

Core Ideas

Talk with a Group

Now, think about the video Sensing Biodiversity from the Sky, which you saw in Get Started. Discuss how what you have learned about solar radiation and Earth's surface can help you understand the warming Earth.

EVALUATE

CONCEPT

4.4

Remote Sensing

Student Objectives

By the end of this lesson:

- [ ] I can develop models of echolocation and sonar that describe how systems and their components interact to achieve proper functioning.
- [ ] I can analyze and interpret data collected by remote sensing technology and evaluate its usefulness in local communities and global communities.
- [ ] I can gather and synthesize information to explain how various methods of remote sensing can be used to monitor stability and change in local environments and worldwide.
- [ ] I can analyze and interpret data from satellite images to argue from evidence that human-induced changes to the environment affect Earth's systems.

Key Vocabulary

aquatic, deforestation, echo, electromagnetic spectrum, environment, glacier, infrared, reflect, satellite, sound wave

Quick Code
ca8828s

Activity 1

Can You Explain?

How can satellite technology be used to monitor human impacts on Earth systems?

Quick Code
ca8829s

Activity 2

Ask Questions

Quick Code ca8830s

Columbia Glacier 1986 and Columbia Glacier 2017

The images you see are of the Columbia Glacier taken 30 years apart. These are digital images that represent sets of data collected by a satellite with sensors that record different wavelengths of reflected and emitted electromagnetic radiation. The data is transmitted to Earth as reliable digital wave pulses and processed by computers. The results are digital images.

Study the images, and then **answer** the questions that follow.

Columbia Glacier 1986

Columbia Glacier 2017

CCC Stability and Change

What differences do you observe between the two images?

Why are images like these useful in determining the extent of changes over large areas?

Why are satellites like the one that collected this data vital tools for monitoring a planet that is undergoing rapid changes due to the impact of an expanding human population?

ENGAGE

Read the text, **observe** the image, and then **answer** the questions that follow.

The image below is one of the satellites built to monitor conditions on Earth. This one has an orbit that enables it to take images of the whole planet once every 16 days. It contains sensors that collect specific wavelengths of electromagnetic radiation that are useful in collecting data on water resources. Other satellites use different wavelengths to collect data on different phenomena.

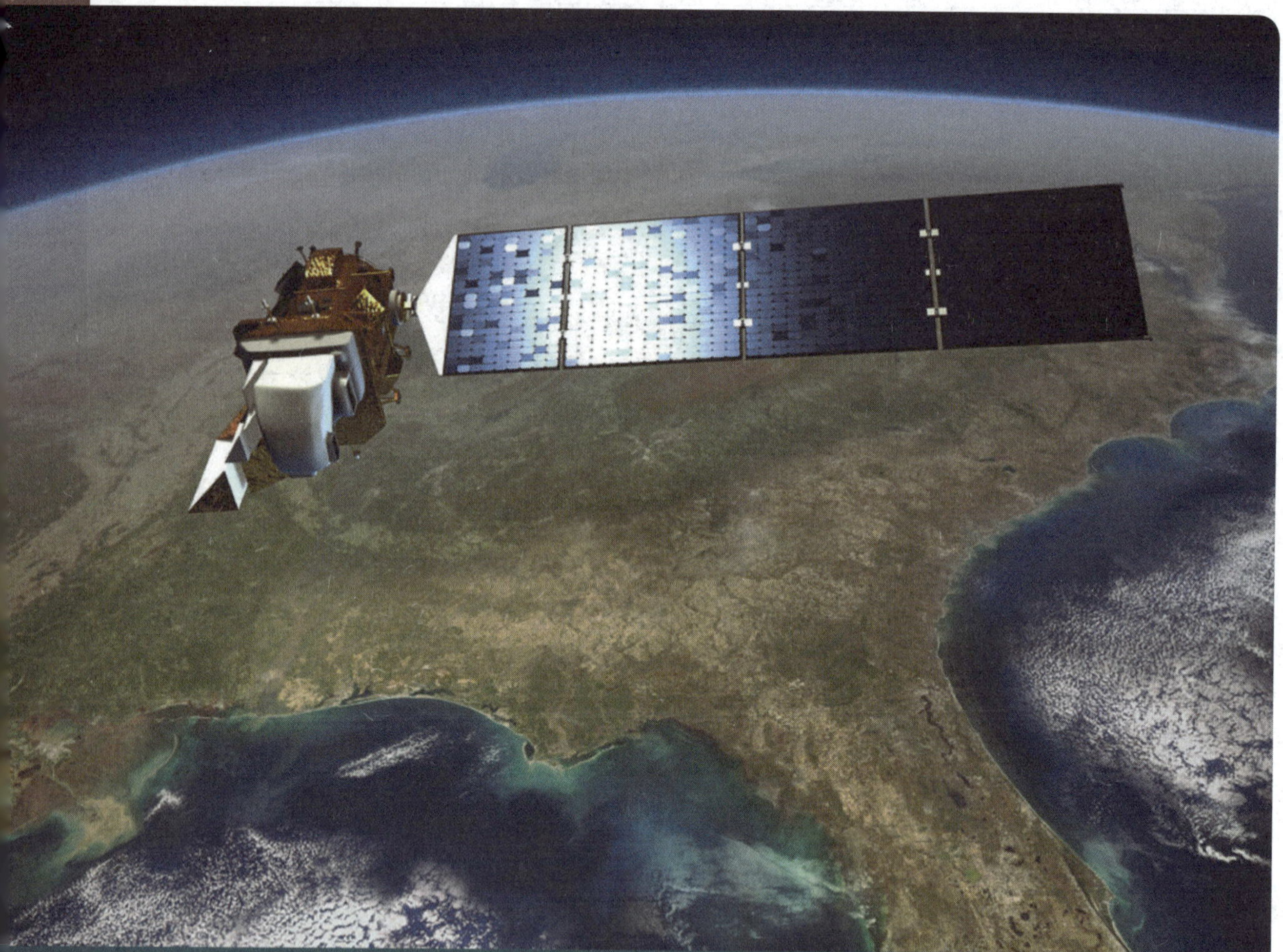

Landsat 8

A satellite like this stays in orbit for many years. How are its instruments powered?

What questions do you have about satellites or the information they give us? Write your questions below.

ENGAGE

Activity 3

Evaluate

Quick Code ca8836s

What Do You Already Know About Remote Sensing?

Changes from Space

Examine the two satellite images showing Trinity Lake Reservoir in California, then **read** the statements and **circle** which inferences can be made from the images. **Circle** all that apply.

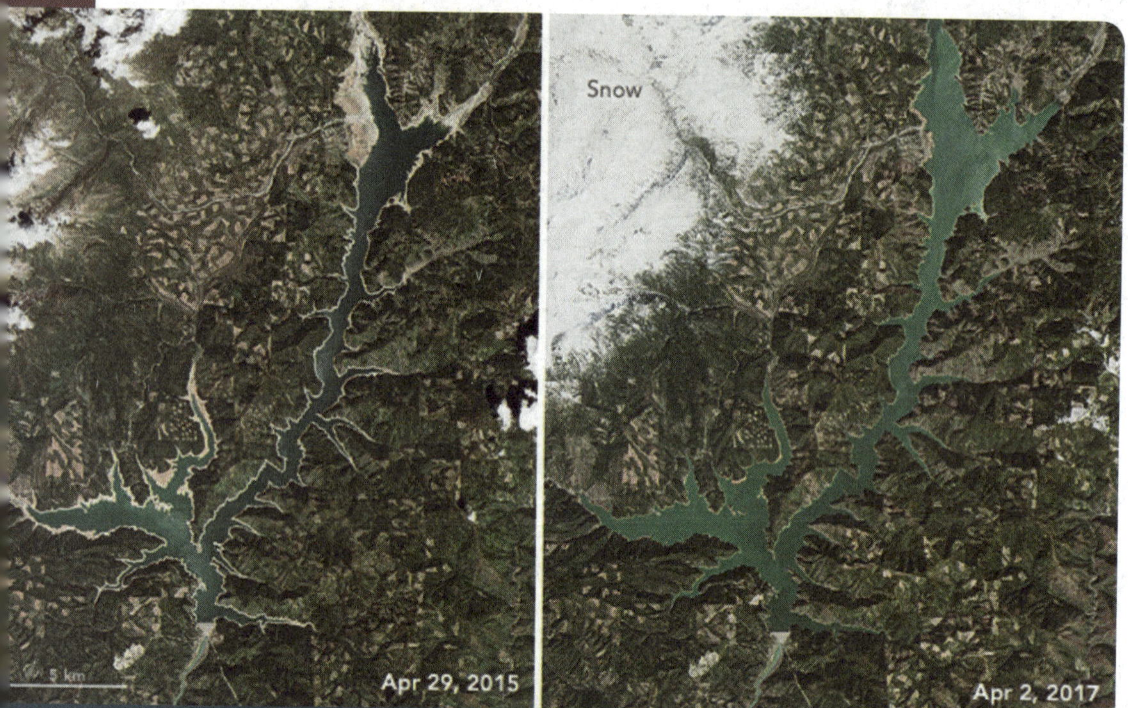

Trinity Lake Reservoir

A There was more water available for use on April 2, 2017, than on April 29, 2015.

B The area shown received snowfall during the months prior to April 2, 2017.

C A large amount of water was released from the reservoir in April 2015.

D The area received no snow during the winter of 2014–2015.

ENGAGE

Using the Global Positioning System (GPS)

Which of the following best explains the benefits of using GPS versus using a traditional map? **Select** your answer.

A GPS is powered by a computer and always accessible.

B GPS works in areas where it has connections to satellites.

C GPS includes the location of major highways.

D GPS accounts for real-time changes in roadways and the natural landscape.

How Are Sound Waves Used to Collect Information About the Environment?

Activity 4

Analyze

Quick Code ca8837s

Remote Sensing: Observing Without Making Contact

Read the following text and then **discuss** each question with three of your peers before writing your response. As you read, **complete** the chart to differentiate remote sensing and standard fieldwork.

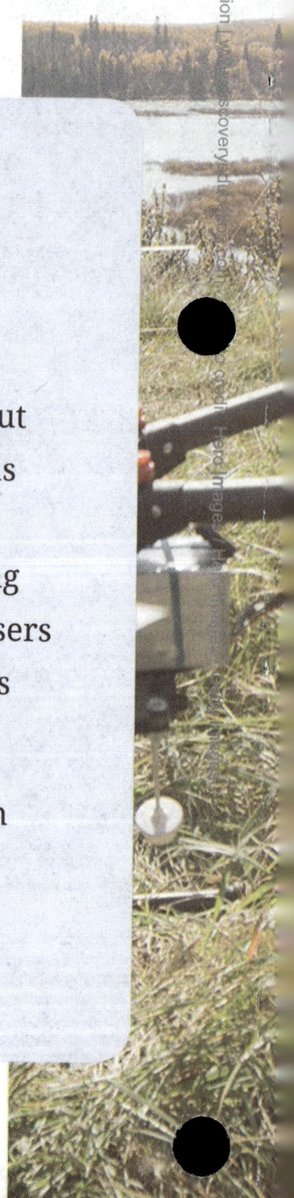

Remote Sensing: Observing Without Making Contact

There are two main types of methods for monitoring changes in Earth's **environment**: field work and remote sensing. Field work involves going out "into the field," by car or on foot, to measure, count, and make observations using tools like tape measures, thermometers, pH meters, and surveying instruments. Remote sensing involves making observations without making contact with an object. Flying over a forest to take photographs or using lasers to measure tree heights is a type of remote sensing. Much remote sensing is done using aircraft as well as satellites that orbit Earth.

Remote sensing can also involve setting out sensors, such as GPS devices, in the field that transmit data to computers far away. Remote sensing devices use waves to gather and transmit information. **Sound waves** and, more commonly, electromagnetic waves, are used for remote sensing.

CCC Cause and Effect

Field Work	Remote Sensing

EXPLORE

Recall your earlier work with echoes, then **answer** the questions below.

What causes an echo?

____

____

____

____

____

____

____

What information does an echo provide about objects in the environment?

____

____

____

____

____

____

____

How do animals use echoes to “remote sense” their environment?

EXPLORE

Activity 5
Evaluate

Quick Code
ca8915s

Bat Echolocation and Echolocation vs. Sonar

Bat Echolocation

Watch the video.

Bat Echolocation

CCC **Cause and Effect**

Use what you learned from the video to **complete** the activity.

A bat uses echolocation to locate four mosquitoes in its local environment. Timing information used in this process is shown in the table below. **Use** this information to **write** each mosquito's number in the correct box in the figure.

Mosquito #	Time for echo to be detected (milliseconds)
1	0.02
2	0.10
3	0.90
4	0.006

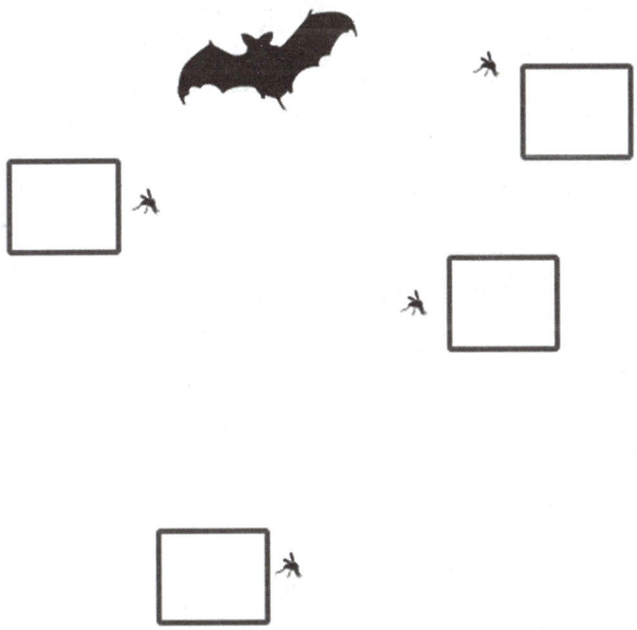

EXPLORE

Echolocation vs. Sonar

Watch the video, then **answer** the questions that follow.

Explain how systems are involved in both echolocation and sonar.

Draw a sketch to show the components of each system.

EXPLORE

Explain how the parts of each system work together to enable the proper functioning of each process.

Compare the two systems by completing the table below.

Similarities	Differences

Echolocation vs. Sonar

Consider the following phrases. Some phrases apply only to echolocation, and some apply only to sonar. Others apply to both echolocation and sonar. **Write** the phrases in the appropriate box in the table.

- uses sound waves
- assists in locating objects
- can be passive or active
- used by animals
- assists in navigating
- developed by humans
- involves reflection of sound waves

Characteristics of Echolocation Only	Characteristics of Sonar Only	Characteristics of Both Echolocation and Sonar

EXPLORE

How Do Satellites Help Scientists Measure and Monitor Earth's Environment?

Activity 6

Observe

Using GPS

Quick Code
ca8916s

Think about how you use GPS for navigation in your car or on your phone, then **answer** the questions that follow.

Where have you encountered GPS?

What does a GPS receiver have to do before it can determine your position?

Watch the video.

Using GPS to Track Wildlife

Use what you learned from the video to **answer** the following questions.

Can you identify the parts of GPS that are needed to enable scientists to track elephants?

How did scientists use information from other remote sensing satellites to help them learn more about elephant distribution?

EXPLORE

Activity 7

Analyze

Remote Sensing Using Satellites

Quick Code
ca8917s

Read the text and **observe** the images, then **answer** the questions that follow.

Remote Sensing Using Satellites

Most remote sensing involves the use of electromagnetic radiation. Electromagnetic waves reflected or emitted by objects provide information about the objects. Most remote sensing of Earth is done using satellites and aircraft. Earth-observing satellites are fitted with sensors that are sensitive to different parts of the **electromagnetic spectrum**. Like the sensors in the camera of your smartphone, these convert electromagnetic radiation into electrical digital signals. These digital signals are then transmitted to receivers on Earth and converted into images by computers. Digital signals are used because they provide a higher quality signal than analog signals and can also carry more information per second.

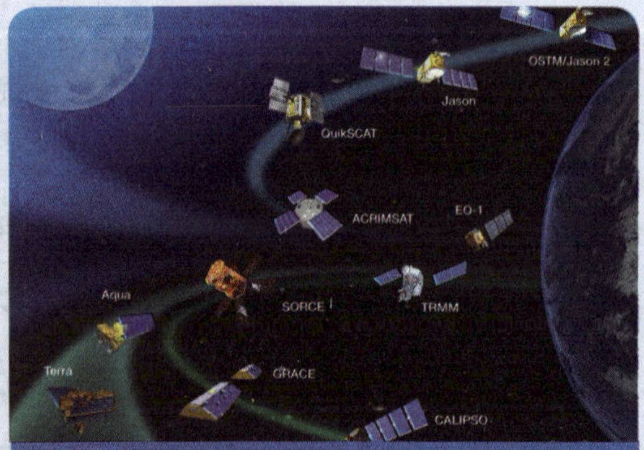

Earth-Observing Satellites

Like sonar, remote sensing using satellites can be active or passive, collecting natural electromagnetic radiation reflected or emitted by objects. Satellites that use active remote sensing emit signals and then detect their reflection. For example, a laser remote sensing satellite projects a laser beam down to Earth's surface and measures the time that it takes for this beam to **reflect** back to its sensor.

Types of Remote Sensing from Satellites

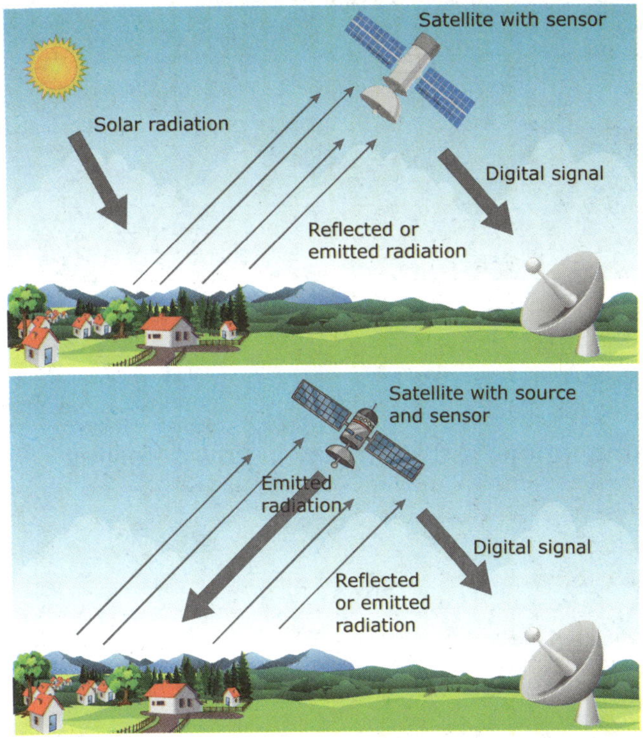

Passive remote sensing

Active remote sensing

How does passive remote sensing occur? How does active remote sensing occur?

What is the difference between active and passive satellites?

What different roles does electromagnetic radiation play in how satellites work?

Use your knowledge of analog and digital signals to **explain** why today's satellites and spacecraft use digital rather than analog signals to communicate with Earth.

EXPLORE

Activity 8

Reason

Using Remote Sensing Data

Quick Code ca8838s

In this activity, you will use global imaging software and remote sensing imagery to examine historical imagery of various places in and around your community to identify changes in land use. You will then compile your results to identify patterns of change.

What materials do you need? (per group)

- Access to computer with global imaging software
- Sticky notes
- Markers

Procedure

Access images of your local community taken recently and in the past using a global imaging database. Print and study the images to find patterns in changes in your community's land use. Determine the factors that caused the changes.

SEP Analyzing and Interpreting Data

SEP Engaging in Argument From Evidence

CCC Stability and Change

Reflect

What pattern(s) do you see in the timeline?

What factors might explain the major changes in your community over the time period studied?

How does having data from all study sites shown together on the timeline help in the analysis?

EXPLORE

How Is Remote Sensing Used to Monitor Long-Term Changes in the Environment?

Activity 9

Observe

CFCs and the Ozone Layer

Quick Code
ca8839s

Watch the video and **examine** the image. With a partner, **answer** the questions that follow.

CFCs and the Ozone Layer

Where did the compounds (CFCs) that entered the atmosphere come from? Why weren't people worried about them at first?

SEP Engaging in Argument from Evidence

CCC Patterns

CCC Stability and Change

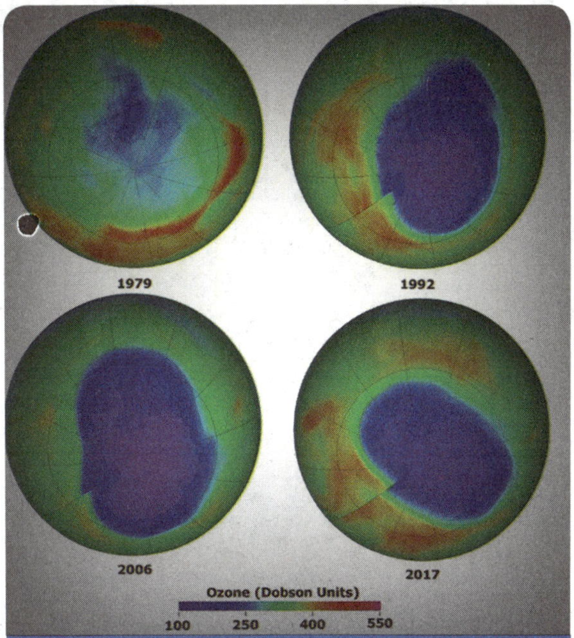

Remote Sensing of the Ozone Hole (1979–2017)

What harm did these compounds cause?

What changes do you observe? How can you account for these changes?

EXPLORE

Activity 10

Analyze

Changes in Landscape and Forest Cover

Quick Code
ca8840s

Read the text and **study** the image. Then, **answer** the questions that follow.

Changes in Landscape and Forest Cover

Remote sensing is useful for showing long-term changes in landscape, water availability, and land use on a regional or national level. As many of these changes are the result of human activities, it is useful for identifying long-term human impacts, including environmental degradation and climate change. In the image, datasets have been combined to provide an image that shows how human activities, including felling, burning, and introducing invasive species, have impacted forest cover.

SEP Engaging in Argument from Evidence

CCC Stability and Change

CCC Patterns

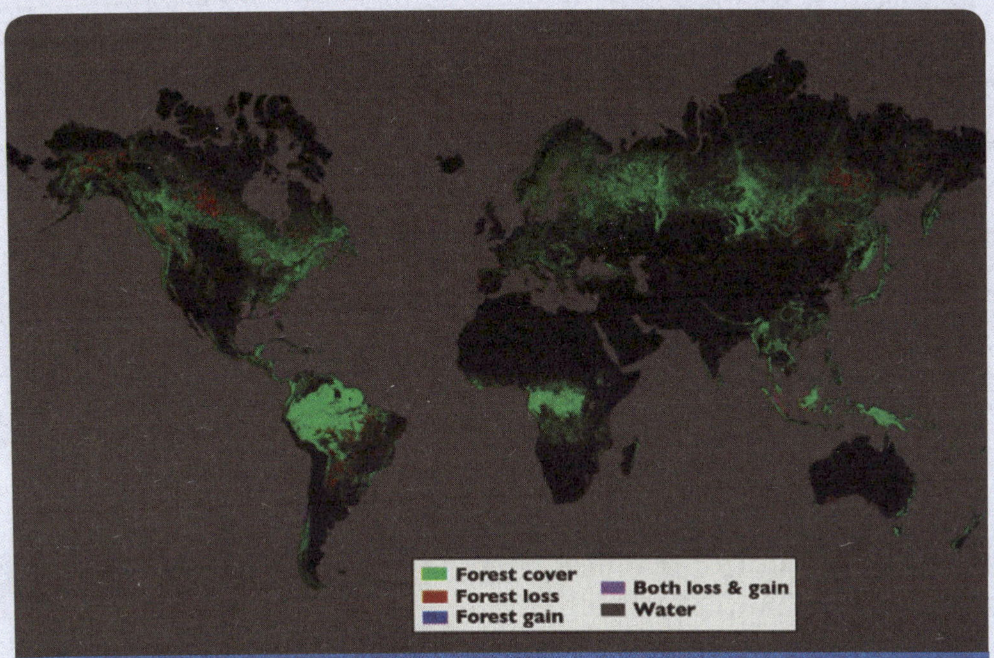

Changes in Global Forest Cover 2000–2012

EXPLORE

What does the image tell you about the distribution of forested areas around the globe and the locations of the areas of major loss?

Think about the uneven distribution of forest loss. Why have some areas lost forest while others have gained them?

Watch the video, then **answer** the question.

How does deforestation cause soil erosion?

Activity 11

Evaluate

Quick Code
ca8919s

Comparing Images: Salton Sea

Read the text, **compare** the images, and then **answer** the question.

Examine the two infrared satellite images from 1984 and 2015 showing the Salton Sea. The Salton Sea was formed accidentally when irrigation canals carrying water from the Colorado River overflowed into the Central Valley of California. Bright red areas indicate agricultural fields. This area is a hotspot for asthma, with rates about twice the regional average.

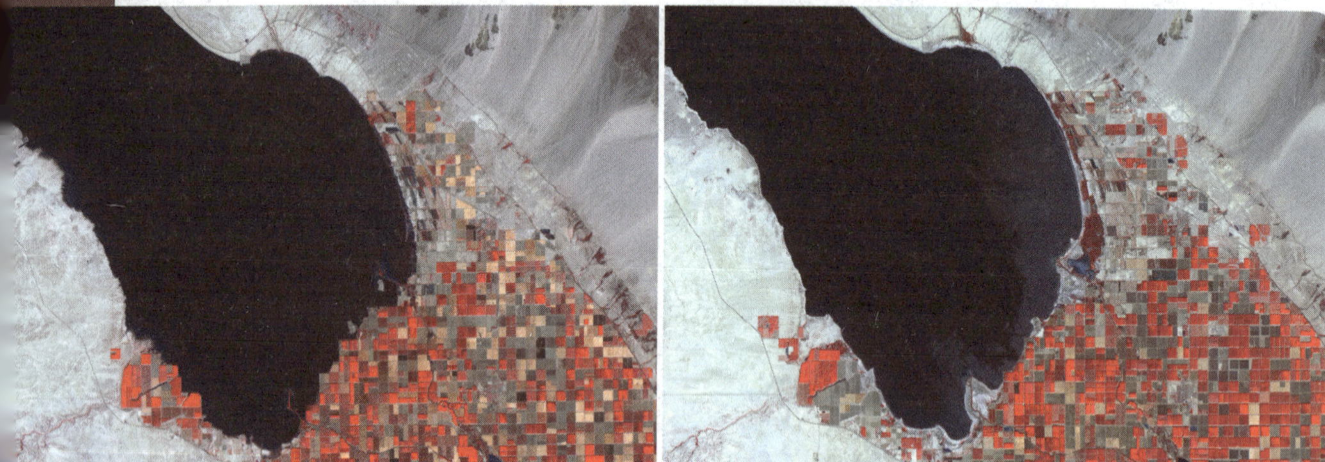

Salton Sea 1984 | Salton Sea 2015

Explain what infrared imagery is, what changes occurred around the Salton Sea between 1984 and 2015, and how those changes might be linked to the high incidence of asthma.

Activity 12

Record Evidence

Quick Code
ca8920s

Columbia Glacier 1986 and Columbia Glacier 2017

As you worked through this lesson, you investigated and gathered evidence about remote sensing. Now, take another look at the images Columbia Glacier 1986 and Columbia Glacier 2017, which you first saw in Engage.

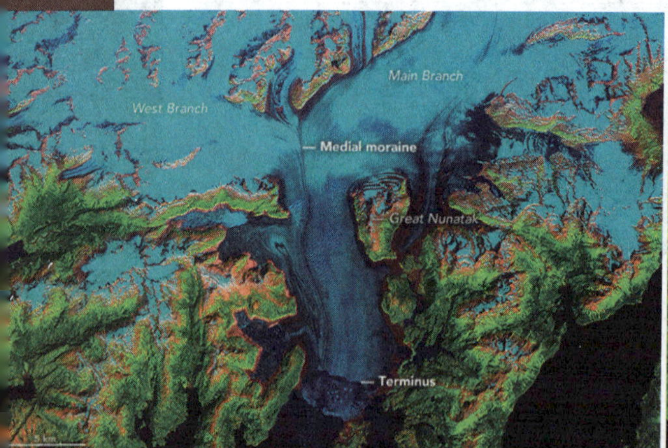

Columbia Glacier 1986

Columbia Glacier 2017

How has your understanding of the use of satellites to monitor changes on Earth changed?

Read the Can You Explain? question from the beginning of this lesson.

Can You Explain?

How can satellite technology be used to monitor human impacts on Earth systems?

Use your new understanding of how satellites are used to monitor changes on Earth to write a scientific explanation answering a question.

SEP **Engaging in Argument from Evidence**

1. **Choose** a question. You may choose to answer the Can You Explain? question or one of the questions you wrote at the start of this lesson.
2. Now, create a scientific explanation to answer the question you chose.

EXPLAIN

STEM in Action

Activity 13

Analyze

Becoming a GIS Specialist

Quick Code
ca8843s

Read the text, and then **complete** the activities that follow.

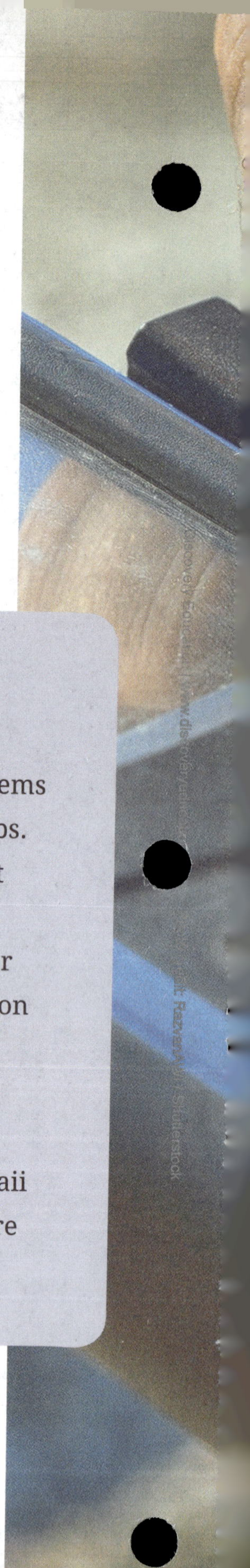

Becoming a GIS Specialist

Have you ever wondered how computer mapping or other navigation systems work? You can use these systems to see roads, restaurants, hotels, and shops. You can then ask to find specific places. They will even calculate the fastest way to get from one place to another. These navigation systems represent one type of geographic information system, or GIS. A GIS is a technology for gathering, managing, analyzing, and mapping data. Any piece of information that is associated with a location can be put into a GIS. GIS analysts bring these data to life to help planning and decision making for a broad array of applications. In the Hawaiian Islands, management of land and water resources is a key issue. Another issue is public safety. GIS analysts in Hawaii can provide critical data that makes planning for the future easier and more effective.

SEP Obtaining, Evaluating, and Communicating Information

Kalisi Fa'anunu Mausio is a geospatial analyst with the Office for Coastal Management, National Ocean Service. Kalisi is a native Hawaiian. She grew up learning to farm and fish for her food. From this background, Kalisi developed a strong bond with nature and describes herself as a champion of natural places. Protecting these places against the ever-increasing pressures of development, consumerism, and population growth is important to Kalisi. As a geospatial analyst, she maps and manages data for many projects. This includes mapping ocean uses, sea level rise, and other issues. She works with coastal managers, local organizations, and government agencies. The data she generates helps in their decision-making processes.

Creating a GIS

You are a GIS analyst and have been asked to help the Hawaii Planning Office. You must determine where public parking for seashore recreation should be located and how large the parking lots should be. You already have a base map of Hawaii. From the list below, **circle** which data layers for Hawaii you will need in your GIS to help solve this problem.

- ☐ population density
- ☐ roads
- ☐ average temperature
- ☐ coastal geography
- ☐ congressional districts
- ☐ land ownership

Justify Your Choices

Choose three data layers and **describe** why each one would be helpful for planning a coastal recreation parking system. You can use layers that you selected earlier, or you can choose others. Be sure to justify your choices.

ELABORATE

Activity 14

Concept Review

Review: Remote Sensing

Quick Code
ca8842s

Now that you have completed the objectives for this concept, **review** the core ideas you have learned. Record some of the core ideas below.

Core Ideas

Talk with a Group

Now, think about the video Sensing Biodiversity from the Sky, which you saw in Get Started. Discuss how what you have learned about remote sensing can help you understand the use of satellites to monitor changes on Earth.

EVALUATE

CONCEPT

4.5

Sustaining Biodiversity

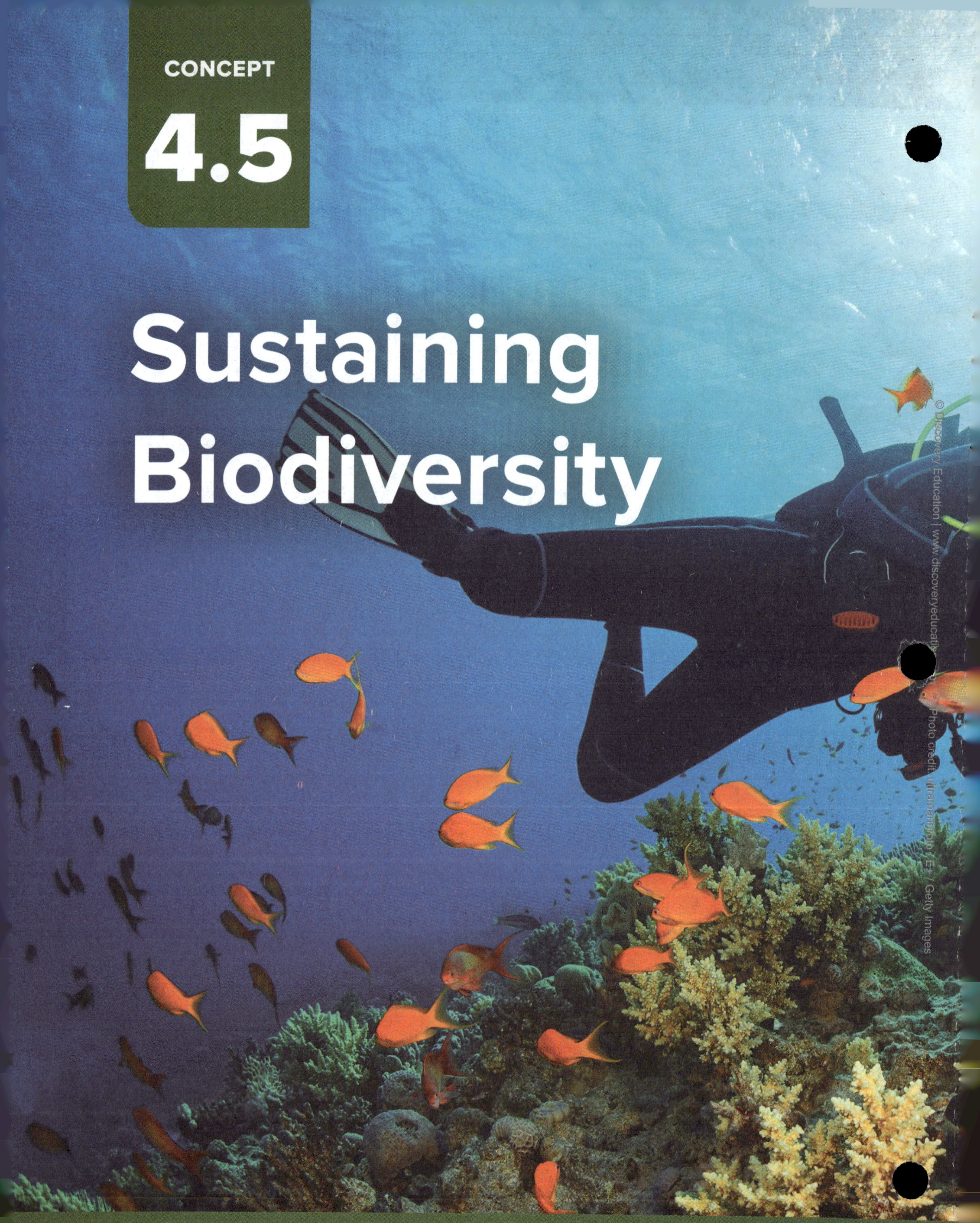

Student Objectives

By the end of this lesson:

- [ ] I can argue from evidence about why biodiversity is important to the stability of Earth's systems and how protecting biodiversity improves life on Earth.
- [ ] I can analyze and interpret data of increased human population size and increased per capita consumption of natural resources and argue from evidence that they affect Earth's systems and local and global biodiversity.
- [ ] I can obtain, evaluate, and communicate strategies for land-use planning and habitat restoration.
- [ ] I can construct explanations of the interconnectedness of ecosystems on Earth and argue from evidence that changes in one ecosystem affect organisms in other ecosystems.
- [ ] I can argue from evidence that climate change is a selective pressure that leads to the increase and decrease of specific traits in populations over time.
- [ ] I can argue from evidence that adding species to or removing species from an ecosystem affects relationships among other organisms both short term and long term.
- [ ] I can use mathematical and computational thinking to evaluate the effectiveness of strategies aimed at restoring endangered species.

Key Vocabulary

biodiversity, biosphere, carbon dioxide, climate, deforestation, ecosystem, endangered species, food chain, food web, fossil fuels, gene, genetic variation, greenhouse gas, keystone species, mangrove, oxygen, pollution, species, sustainable, wetland

Quick Code
ca8850s

Activity 1

Can You Explain?

What human activities harm biodiversity, and what human activities help to sustain biodiversity?

Quick Code
ca8851s

Activity 2

Ask Questions

The Eye of the Condor

Quick Code
ca8852s

Study this image.

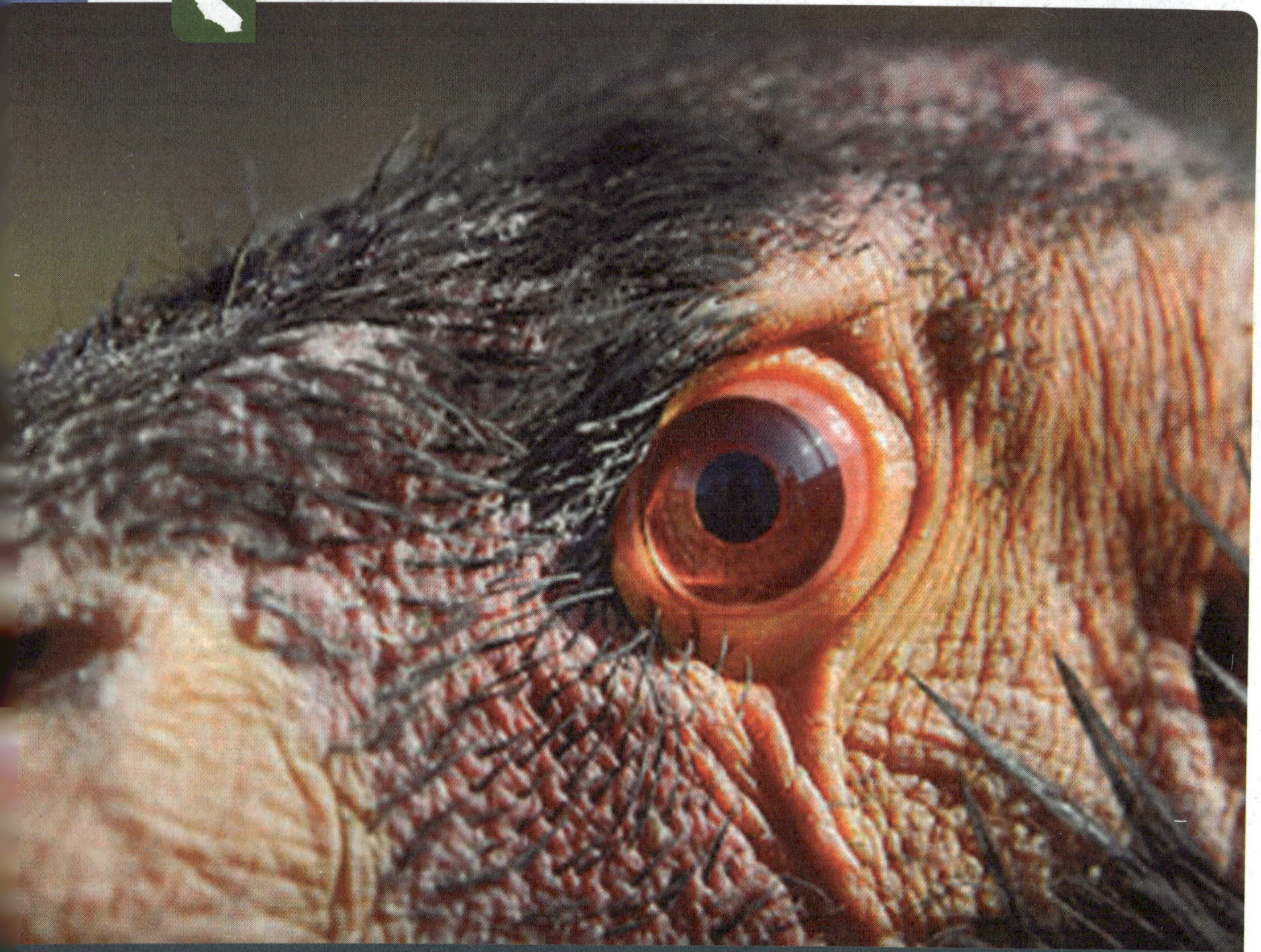

Let's Investigate the Eye of the Condor

ENGAGE

What are some things you already know about the California condor? What are some things you would like to know? How can you find out the things you would like to know? **Use** the chart to **answer** these questions.

What I Know

What I Would Like to Know

How Can I Find Out?

Activity 3
Analyze

Back from the Brink

Quick Code
ca8853s

Read the text. **Highlight** the text in different colors to identify the following:

- The causes of the decline in the condor population
- The actions taken to prevent the remaining population from becoming extinct

Back from the Brink

How many species of birds can you identify? Would you recognize a mockingbird or a scrub jay? What about a turkey vulture, or perhaps a condor? You would certainly be lucky to see a condor. They are very rare, but if you see one, you will know. Although they are very rare, they are also very big—the biggest bird in North America and one of the largest flying birds in the world. Thousands of years ago, condors were seen over much of North America—from the southwest United States, to parts of the East Coast and as far north as British Columbia. Today, they are restricted to the Southwest. Why are these birds so rare?

Condors are nature's trash collectors. They eat carrion—dead animals—which they spot from high in the air as they glide on their three-meter wingspans. As more people populated the West, they shot, poisoned, and trapped these birds and collected their eggs. They also began wiping out the condors' natural food supply as they hunted the animals—elk and antelope—that the condors relied on for food.

Later, pesticides and pollution took a heavy toll, and the trees that condors nested in were felled for timber. There were so many problems facing the condor that the species was not going to survive in the wild. So, in the 1980s, a major conservation project was started to try to prevent them from becoming extinct. Many special studies were conducted. Radio transmitters were placed on the wings of some condors to track them. Wild eggs were collected and hatched at zoos, and this worked to increase the population. Some birds were taken to zoos for captive breeding. Eventually, to prevent their extinction, all the remaining condors in the North America were captured and placed in zoos. The bird is a slow breeder, but by the 1990s, scientists were ready to introduce them back into the wild. The newly released condors were fitted with tags and radio transmitters and carefully released. Many died, but gradually their numbers increased, and scientists tracked their progress. Now, they are breeding again. There are about 400 living in the wild. In some ways, condors were lucky. Although they are still an **endangered species**, they were saved from extinction. The story of their decline is one of many illustrating how human impacts came together to form conditions that made it impossible for a species to survive. In this concept, you will learn about why and how human activities can both harm and protect the environment.

ENGAGE

Use your highlighted text to **fill in** the provided T-Chart with your findings. Then, **answer** the questions that follow.

The California Condor

Causes of Population Decline	Actions Taken to Prevent Extinction

ENGAGE

Did the California condor become endangered because of one or many causes?

Can you think of any other animals that have become endangered? What caused a decline in their populations? Was it one factor or many factors?

Activity 4

Evaluate

What Do You Already Know About Sustaining Biodiversity?

Quick Code ca8854s

Population Problems

Answer the following questions.

Human population growth is a concern in some other countries, but it does not really affect people in the United States. Is this true? **Explain** your reasoning.

How does population growth in other countries impact the availability of food products worldwide, including many imported to the United States?

ENGAGE

Is it possible for a population to run out of resources such as land and food, even if it grows slowly?

Can the U.S. population continue to grow indefinitely?

World Population

Select the statement that correctly describes human population growth over time.

☐ The human population has been growing at the same steady rate for thousands of years.

☐ There had been no growth in the human population until the last 200 years.

☐ The human population had grown rapidly, but it has peaked and is now slowly declining.

☐ The human population has been growing, but the rate of growth has increased greatly during the last 200 years.

Causes of Growth

Consider your answer to the previous question. What likely has caused the trend you identified? **Write** your answer in the space provided.

____

____

____

____

____

____

ENGAGE

Where It Grows

Which statements correctly describe human population growth? **Select** all that apply.

- ☐ Only less-developed countries experience population growth.
- ☐ The United States is experiencing population growth.
- ☐ The population of the United States is declining.
- ☐ Population is growing in most countries but is declining in some.
- ☐ Only very developed countries experience population growth.
- ☐ The population of the United States is unchanging.

What Is Biodiversity, and Why Is It Important?

Activity 5

Reason

Micro-Bioblitz

Quick Code ca8855s

In this investigation, you will work in groups to identify as many different types of organisms in the schoolyard as you can. You will use the cameras on your phones to select species and share them with classmates when you get back to the classroom.

What materials do you need? (per group)

- Paper
- Pencils
- Digital cameras or smartphones
- Sketch pads, drawing media

SEP Analyzing and Interpreting Data

Procedure

Record as many living species as you can find in your schoolyard. You should gather pictures and/or draw each species you find and record the following information:

- Where the species was found
- The type of species
- What the specimen was doing

One way to divide the work could be to have one student find the new species, one student take a picture or draw, and the third student record the needed information about each species. **Record** your findings in the space provided.

EXPLORE

Reflect

What information does a bioblitz give you?

Did the bioblitz give a measure of the biodiversity of the schoolyard?

What are in the limitations of this process?

Do you think you found all the organisms that live there? What if you had a longer time period to complete the activity?

How do you think this information would help scientists?

What would you expect to see during a different season? Different location? Different time of day?

Would it have been better if you had experts to identify all the organisms you found?

EXPLORE

Activity 6

Analyze

Biodiversity and Its Importance

Quick Code
ca8856s

Read the following text and **view** the images and videos. Then, **answer** the questions.

Biodiversity and Its Importance

Biological diversity, often shortened to **biodiversity**, is the variety of life found in a certain area. Global biodiversity is the variety of life found on Earth. Earth's biodiversity comprises all its **genetic variation**, **species**, and ecosystems. Scientists measure biodiversity at different levels of biological organization. First, they can measure the amount of variation in the genes of a species. Populations with more genetic variation are more likely to have offspring that possess beneficial mutations. Such mutations can help species survive and reproduce despite environmental changes. Second, scientists can measure species diversity in a given area or over time. Because species interact, a thriving community with many different species indicates a healthy **ecosystem**. Third, scientists want to understand ecosystem diversity. Each type of ecosystem supports species adapted to its conditions. The number and variety of ecosystems allow Earth's **biosphere** to support millions of species.

Video

Biodiversity

CCC Cause and Effect

EXPLORE

One way to measure biodiversity is to count the number of species in an area. When comparing areas of the same size, some have far more species than others. It is impossible to exactly determine the number of species in an ecosystem because organisms move and ecosystems change over time. Some species are very tiny, and others have not yet been identified by scientists. Most scientists count the number of organisms they can find in a few sample areas within an ecosystem and use this information to calculate and compare biodiversity.

Species are not spread uniformly over Earth's surface. Some parts of the planet, such as the tropics, have many more species than others. Knowing which organisms are in an area is important because it can help scientists identify those areas that are particularly rich in species. These hotspots of species richness can then be prioritized for monitoring and conservation. Biodiversity can be used to measure the health of ecosystems. Loss in biodiversity is one indicator that an ecosystem may be under threat.

Scientists study species diversity because they want to understand how it is affected by changes in ecosystems. If an ecosystem is damaged, high species diversity can allow it to recover quickly. Ecosystems provide many valuable functions, or services, that benefit humans.

Video

Services Provided by Wetlands

Biodiversity and Its Importance *cont'd*

For example, coral reefs, **mangrove** forests, and salt marshes provide spawning areas for fisheries and reduce damage by storms to coastal areas. Forests absorb **carbon dioxide** produced by the burning of **fossil fuels**, produce **oxygen**, and provide a wide range of products from lumber to food and medicine. Freshwater wetlands play a vital role in reducing flooding and purifying water. When biodiversity is lost, ecosystems are less able to provide these valuable services. Biodiversity directly sustains human life. We depend on many plant and animal species for food. In agriculture, many wild animals serve as natural pest controllers and pollinators. As the environment changes, biodiversity provides a source of genes that may solve problems related to food, energy production, and medicines. There are rare plants and animals that may hold cures to human diseases. Biodiversity meets human needs, but people also benefit from the beauty and inspiration provided by Earth's variety of life forms. Think of the enjoyment we get from hiking in the wild, birdwatching, or watching nature films.

Why is it important to identify species found in an ecosystem?

____

____

____

____

____

EXPLORE

Why would humans be worse off if most of Earth's biodiversity were eliminated?

How Do Human Population Growth and Resource Consumption Impact Biodiversity?

Activity 7

Interpret Data

Human Populations and Biodiversity

Quick Code
ca8857s

Observe the graph, and then **answer** the question.

Human Population Size

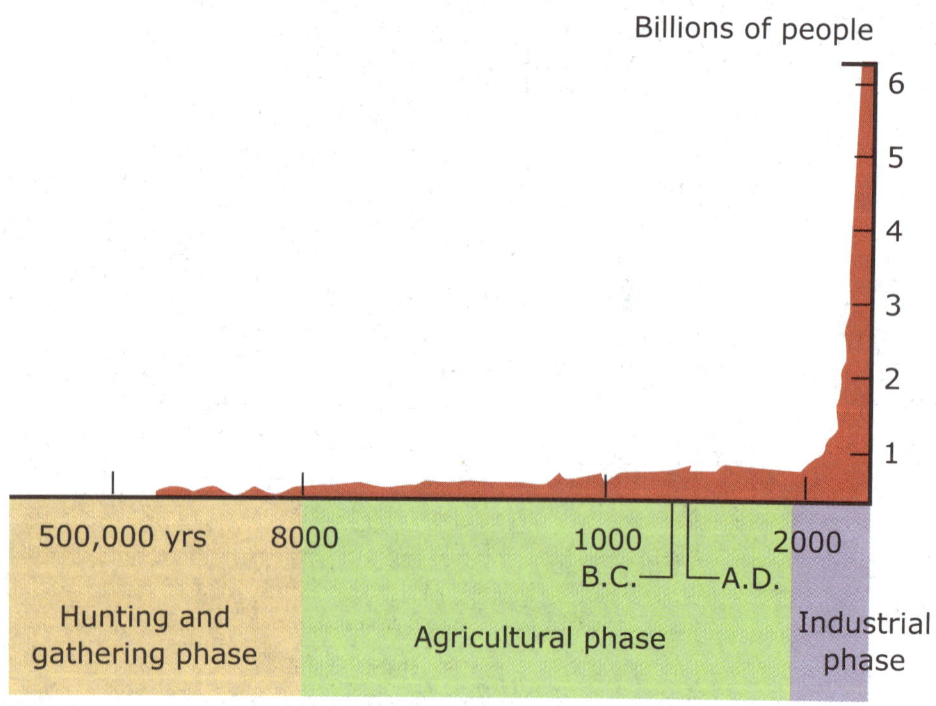

Do you think human populations can continue to grow indefinitely? If not, why not?

____

SEP Analyzing and Interpreting Data

CCC Cause and Effect

Observe the graph, and then **answer** the questions.

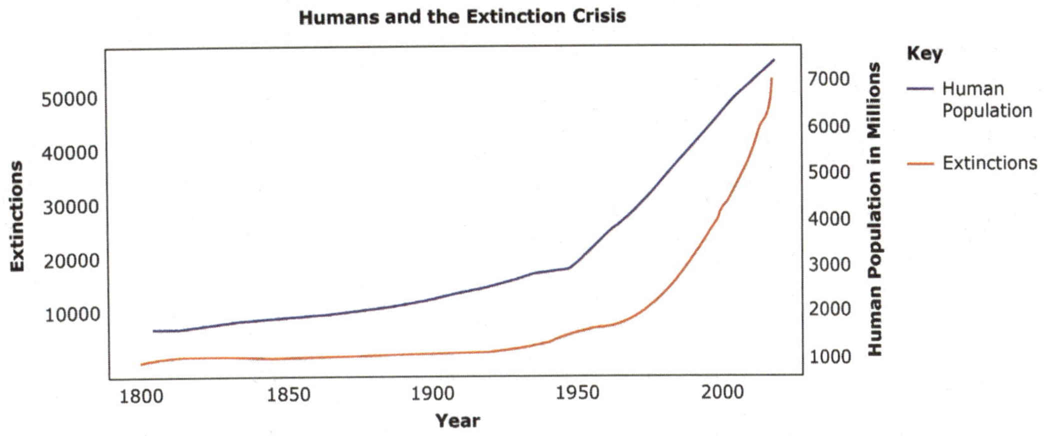

What is the relationship between the rate of human population increase and the rate of extinctions?

EXPLORE

What will happen to the biodiversity of Earth as the human population continues to grow?

Think back to the decline of the California condor. **Answer** the following questions in the space provided.

Do you think the human population increase in California contributed to the decline of the California condor? If so, why?

What would have happened to the California condor if wildlife biologists had not intervened?

Photo credit: ultramarinfoto / E+ / Getty Images

Activity 8

Analyze

Per Capita Consumption

Quick Code
ca8859s

Read the text and **highlight** information in the text that describes how increased consumption by each person damages the biosphere. **Answer** the questions that follow.

EXPLORE

Per Capita Consumption

The number of human beings living on Earth doubles about once every 50 years. Populations grow for many reasons, such as improvements in medical care that allow people to live longer and more babies to survive infancy. Population growth is defined as the number of births exceeding the number of deaths. In 2018, there were about 7.7 billion people living on the planet. In the future, we will need to provide more services to more people, including health care, sanitation, and housing. As our population grows, more and more resources are required to meet the expanding population. Not only are there more people on Earth every year, but each of those people, on average, is consuming more resources than the previous year.

Increased consumption has many bad effects on the environment. By exploiting natural resources and expanding our needs for space, human activities damage natural ecosystems. This reduces biodiversity by causing the decline and extinction of some species. Human consumption damages the biosphere overall and leads to

SEP Analyzing and Interpreting Data

CCC Cause and Effect

Per Capita Consumption *cont'd*

changes in ecosystems. As ecosystems change, the organisms within them are subject to changes in their environment that influence the process of natural selection. Given time, some organisms may adapt to these new conditions; however, usually the changes happen too quickly for adaptation to occur.

Currently, there is a rapid decrease in biodiversity worldwide with extinction rates comparable to those that occurred during earlier mass extinctions in Earth's history. The exact number of extinctions is difficult to calculate because many, perhaps most, species have not yet been discovered. However, scientists conservatively estimate that around 2,000 species become extinct each year.

Do people in poorer countries have the same per capita consumption as people in the United States?

____

____

____

____

____

____

EXPLORE

Based on your answer, do people in poorer countries have the same impact on the environment and biodiversity as people in the United States?

What would happen to Earth's biodiversity if everybody had the same per capita income as people in the United States?

Activity 9

Evaluate

Quick Code ca8860s

Comparing Nations, Energy and Biodiversity, and Suggestions for Change

Comparing Nations

Per capita electricity consumption is a useful indicator of resource consumption in different countries. The table below shows the population sizes and per capita energy consumption in five of the world's nations. **Use** the table to **complete** the activities that follow.

Nation	Population Size	kWh per capita
Bangladesh	157,000,000	310
Iceland	300,000	53,832
Germany	80,600,000	7,035
Qatar	2,200,000	15,309
United States	316,000,000	12,987

SEP Analyzing and Interpreting Data

EXPLORE

Use information from the table to **match** the correct nation to each description below. Some countries may match more than one description, while others may not match any.

Bangladesh	Largest Population
Germany	Smallest Population
Iceland	Lowest per Capita Electricity Consumption
Qatar	Lowest Total Electricity Consumption
United States	Highest per Capita Electricity Consumption
	Highest Total Electricity Consumption

Energy and Biodiversity

Identify the countries with the highest and lowest total electricity usage. What would happen to biodiversity if the population doubled in the country with the highest total electricity consumption? **Compare** this to the effect on biodiversity if the population of the country with the lowest total energy consumption doubled.

Suggestions for Change

What suggestions would you make to the country with the highest electricity consumption to help reduce its impact on the global climate?

EXPLORE

How Can Land-Use Planning and Habitat Restoration Help Protect Biodiversity?

Activity 10

Analyze

Quick Code
ca8877s

What Is Land-Use Planning?

Think back to the Using Remote Sensing Data activity that you completed in the last concept. **Answer** the following questions based on that activity.

When you were observing the images, what different types of land use did you observe?

____

____

As the landscape changed around the communities, how did land use change?

____

____

____

What impacts could these changes have had on the animals and plants living in the area?

____

____

____

CCC **Cause and Effect**

CCC **Stability and Change**

As you read the following text, **highlight** the different ways that land-use planning can be used to protect biodiversity. **Record** the different ways in the space provided.

EXPLORE

What Is Land-Use Planning?

Habitat destruction is widespread, but there are ways to prevent it. One of the most important is careful land-use planning. Most countries, states, or cities have development laws that regulate land use and require careful research before development permission is given. Many restrict development of certain ecosystems (such as wetlands) and areas that are habitats for threatened or endangered species. They may set aside open spaces to preserve some of the existing habitat. Areas such as nature reserves, wildlife refuges, and state and national parks are set aside as no-go areas for development or exploitation. Most human activities damage natural ecosystems, but humans can fix some of the problems they have created. By carefully studying and managing ecosystems, conservation scientists can sometimes restore them to their original states.

Habitat destruction is widespread, but there are ways to prevent it. Most cities have laws that regulate land usage. Many set aside open spaces to preserve habitat. Cities usually specify the locations where housing, public spaces, factories, and landfills can be developed.

What Is Land-Use Planning? *cont'd*

Citizens can hold their town or city accountable for responsible growth and development. To help protect their neighborhoods and natural resources, people can phone, write letters, or petition companies and government agencies to consider the environmental effects of their actions.

How can land-use planning be used to protect biodiversity?

EXPLORE

Watch the following videos and **think** about the impacts human activities can have on biodiversity. How can biologists and engineers use land-use planning and innovative technologies to solve land-use conflicts? **Use** your thoughts from the videos to **fill in** the provided problem/solution graphic organizer.

Growing Crops in Cities

Freshkills Landfill

What human activities harm biodiversity, and what human activities help to sustain biodiversity?

Problem		Solution
	→	
	→	
	→	

How Can Monitoring and Reducing Pollution Help Conserve Biodiversity?

EXPLORE

Activity 11

Observe

Quick Code
ca8863s

Treating Waste Water

What is meant by the term *pollution*? **Work** in a group of four to **brainstorm** different types of pollution, sources of pollution, and ways each type of pollution might be prevented. **Record** your ideas in the chart.

Type of Pollution	Sources	How It Could Be Prevented

CCC Cause and Effect

Watch the following video to learn about one specific example of preventing pollution: treating waste water.

Treating Waste Water

Read the following text. **Look** for the ways scientists can track pollution in a system.

Scientists can use biodiversity to monitor the health of an ecosystem. For example, by tracking the types and numbers of organisms they can determine pollution levels. Healthy unpolluted freshwater ecosystems, such as streams and rivers, contain a wide diversity of animals. Some of these are more sensitive to different types of pollution than others—they are biological indicators of pollution. The presence or absence of these organisms can therefore be a useful measure of water pollution. By using biological indicator species, together with chemical testing, scientists can make informed decisions about laws that affect how much pollution is allowed.

Activity 12

Reason

Quick Code
ca8864s

How Can We Reduce Pollution?

In this investigation, you will work in small groups to survey the school and come up with a list of ways to possibly reduce pollution. You will consider the ideas as a class and vote on the ideas that are likely to be most easily implemented. A delegation will be selected to present the ideas to the principal.

Procedure

Work in small groups to survey students and teachers about the ways your school could reduce pollution. As you conduct your survey, keep the focus on methods that reuse or recycle materials. Record the suggestions you are given and keep a tally of how many people suggest each method. Once you have finished your survey, report your findings to the class and discuss the ideas that came out of the survey.

If a survey is not an option, tour the school as a class. While on the tour, record your ideas to reduce pollution in your school. When you return to the classroom, record your ideas on the board.

What were some ideas that your class came up with either through a survey or a tour?

____

____

____

____

____

____

____

CCC Cause and Effect

EXPLORE

What materials do you need? (per group)

- Pencils
- Poster board
- Markers

As a class, vote on which ideas you feel would be most easily implemented. Create a poster that lists the ideas that received the most votes. Your poster should include information about the amount of interest shown school-wide (or classroom-wide) as well as the measurable evidence of impact of each suggestion. **Record** a copy of your class's poster in the space provided.

Reflect

Which suggestions were surprising?

How do you think this type of survey helps reduce pollution?

What do you think are some limitations to this survey?

Activity 13

Observe

Climate Change and Biodiversity

Quick Code ca8865s

What do you already know about the causes and nature of human-induced global climate change? **Answer** the following questions in the provided spaces.

What are the major causes of climate change?

____

____

____

____

How is the climate changing?

____

____

____

____

What effects of climate change have you observed?

____

____

____

____

CCC Stability and Change

CCC Cause and Effect

Photo credit: ultramarinfoto / E+ / Getty Images

Watch this video. **Record** the impacts that climate change can have. **Organize** your ideas in the graphic organizer provided

Impacts of Climate Change

Impacts on Human Activities	Impacts on Natural Habitats and Wildlife

EXPLORE

Activity 14

Analyze

Adapting to Climate Change

Quick Code
ca8867s

Read this passage to analyze how climate change can affect genetic variation and natural selection. Then, **answer** the questions that follow.

Adapting to Climate Change

Human activities, such as the burning of fossil fuels and deforestation, have caused an increase in greenhouse gases in the atmosphere. These gases trap heat radiating from Earth toward space and prevent it from escaping. The heat remains near Earth, causing an overall increase in temperature on Earth's surface. This climate change caused by the actions of humans impacts all living things on Earth.

Evidence is surfacing that salmon are adapting to climate change.

CCC Stability and Change

CCC Cause and Effect

In addition to glaciers melting and shorter winters in many locations, habitats are changing all over Earth. In many cases, the changes are so significant that species must adapt or become endangered or even extinct.

One particularly concerning aspect of global temperature change is that it is happening relatively quickly. As a result, adaptation has to happen relatively quickly as well. Species must evolve over a shorter period of time to survive.

One study with brewer's yeast showed that a shortened period of evolution is possible. While evolutionary changes have taken hundreds of thousands of generations, adaptive changes in the genes of brewer's yeast were observed to occur in just 25 generations. Researchers introduced salt to a yeast population, causing stress that reduced growth and ultimately caused death in many yeast cells. Although initial results showed a sharp decrease in the number of cells, the population eventually stabilized. The cell density, or number of cells in a given volume, began at 100 percent. It then dropped to 10 percent after the salt was introduced. Over the course of about 50 hours, the cell density began to increase again. It returned to 100 percent within 125 hours. Researchers believe this happened because a small number of resistant cells were able to survive in a high-salt environment. As these resistant cells reproduced and became more numerous, the yeast population rebounded.

Adapting to Climate Change *cont'd*

So how does this relate to organisms stressed by higher temperatures? It means that researchers may be able to observe some rapid evolutionary changes occurring on Earth. Some coral species are very sensitive to the changes in temperature. When corals are heat-stressed, they expel the algae that live inside and give them their color. This is called bleaching. Bleaching can be lethal because the algae provide a source of much-needed nutrients to the coral. One coral species, the table coral, seems to have adapted to warmer waters, becoming more resistant to bleaching. Like the brewer's yeast, some coral had genetic material that made them resistant. These corals survived and reproduced, causing changes in the species through natural selection.

Tawny owls are another species that appear to be undergoing evolution due to warmer temperatures. The tawny owl is a medium-sized owl found in woodlands across Europe and Asia.

Its color varies in the amount of brown that is present. Researchers believe that changes in color patterns on tawny owls may be due to natural selection. These owls live and hunt in regions that are often covered by snow. Populations of owls in areas with more snow tend to have more light-colored owls. Their coloring allows them to blend in with the snow. As winters become milder due to increasing global temperatures, there is less snow cover. In regions with less snow cover, researchers have found an increase in the number of owls with more brown coloring. The greater amount of brown helps the owls blend in with the bare groundcover. The shift toward greater numbers of brown owls suggests that the owls are adapting to the new conditions.

Salmon provide yet another example of rapid adaptation to environmental change. Salmon swim from the ocean into freshwater streams to produce offspring. Scientists have found that this drive to swim upstream to spawn is a genetic trait. However, a study of migratory and genetic data for the last 40 years shows that changes have occurred that coincide with the shortening of winter each year. Salmon that have a genetic marker that causes them to migrate earlier have gradually increased. This increase took place beginning in the 1980s when the late migrating genetic marker made up only 10 percent of the salmon population. The water temperature increased an average of 1°C between the 1980s and 2011. Scientists believe that the change in the salmon population is an adaptation to climate change. It could be that the salmon that migrate later in the season will continue to decline to the point where they will no longer exist.

EXPLORE

What role do humans play in the need for species to adapt to warmer temperatures?

How would climate change affect species in a colder climate that has seen a one-degree rise in temperature?

Why would late-migrating salmon be in danger of extinction?

How Can Adding Species To and Removing Species From Habitats Impact Biodiversity?

Activity 15

Analyze

Quick Code
ca8868s

Hunting, Overexploitation, and Biodiversity

What would happen if a single species were removed from an ecosystem? **Work** in a small group to **predict** what would happen if one species of your choosing was removed from an ecosystem of your choosing. After brainstorming, **answer** the questions below. **Share** your answers with the class and **evaluate** the answers of your classmates.

How do you think other species would be affected?

____

____

____

____

____

____

CCC **Patterns**

CCC **Stability and Change**

CCC **Cause and Effect**

What might be the short- or long-term consequences of such a loss?

____

____

____

____

____

____

Read the following text, focusing on the impact of fishing (a form of hunting) on biodiversity. **Use** the graphic organizer to **record** the impacts of certain practices and measures that could be used to prevent or reduce the impacts.

EXPLORE

Hunting, Overexploitation, and Biodiversity

All human societies, past and present, have reduced biodiversity, and much of this reduction has come from hunting. For example, when humans first arrived in North America, there were many strange species of mammals that were rapidly hunted to extinction.

Hunting is driving some species to extinction today, although hunting is often just one contributing factor. The populations of many species are much diminished by hunting, particularly in the tropics where animals are hunted for bushmeat (the meat of African wild animals). In the United States, many predators have been hunted to the point that they are no longer present in ecosystems. The mountain lion and gray wolf are two species that inhabit only a tiny part of their original range. Hunting is a problem when it removes species that play a major role in keeping ecosystems healthy. These species are called **keystone species**. For example, the beaver is a keystone species because of the way it modifies and creates wetlands that are home to many other species.

Carefully monitored and controlled hunting is **sustainable** and is widely practiced, particularly in the hunting of deer and some species of wildfowl. In many areas, hunting is essential to keep the populations of some animals in check. This is particularly true where predators, such as wolves, have been eliminated.

Hunting is carried out on an industrial scale in the world's oceans in the form of fishing. Fish make up around one-fifth of the protein eaten by humans. Some of these fish come from fish farming, but most are caught in the wild. In some parts of the world, freshwater fishing is important; however, most fishing takes place in the oceans.

EXPLORE

Over the past 50 years, fishing has become industrialized, and many traditional fishing communities find themselves competing with large ships using very efficient fishing techniques. These techniques commonly lead to overfishing. If nets with a fine mesh are used, young fish are caught, reducing the future breeding population. Year-round fishing may catch fish during important spawning seasons. These practices, combined with fishing in breeding areas, diminish fish stocks and reduce biodiversity. Fisheries all around the world are showing signs of overfishing. Current rates of fishing are not sustainable.

The impact of overfishing goes beyond the removal of food species. Fishing techniques generate bycatch, a term used to describe unwanted organisms that are netted or hooked in error. Bycatch includes unwanted fish species, seabirds such as albatrosses, turtles, and marine mammals like dolphins.

All countries with a coastline have areas over which they control fishing and have regulations determining the methods of fishing allowed and the seasons when it can be carried out. Many of these regulations are designed to reduce bycatch. Some countries, such as Canada and the United States, carefully monitor fishing in their waters. By regulating fishing zones, net design, and mesh size; limiting fishing seasons; and imposing catch quotas, governments have helped fishing stocks recover. But in international waters outside these limits, there are no such regulations.

Aquaculture is becoming an important source of seafood such as prawns, shrimp, mussels, and oysters. This involves farming **aquatic** organisms under controlled conditions. These organisms may be farmed in ponds or in cages in lakes or the sea. Aquaculture reduces pressure on some natural ecosystems, but it can reduce biodiversity because if farmed fish escape into natural habitats, they can become invasive. Mangrove habitats, vital for the spawning of many species of fish, may also get converted into shrimp farms.

EXPLORE

Practices	Impacts	Preventative Measures
Using large “industrial” fishing vessels		
The use of nets with fine mesh		
Year-round fishing		
Fishing in breeding areas		
Bycatch		

Activity 16

Observe

Something Fishy

Quick Code
ca8869s

What does the term *sustainable* mean? What would you expect from a sustainable fishery?

Go online to explore how different rates of harvest can result in sustainable or unsustainable fishing practices. Then, **answer** the questions that follow.

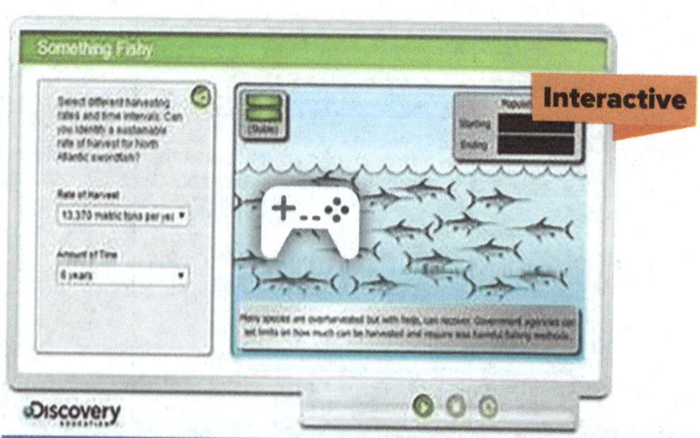

Something Fishy

Define *overharvesting.*

What would happen if we continued to overharvest swordfish?

Which rate of harvest is twice the sustainable rate?

What can governmental agencies do to help prevent overharvesting?

Work in small groups to **discuss** the following issue. **Record** the ideas you come up with below.

A major international scientific study released in the journal *Science* predicts that if current trends in the decline of fish stocks continue, all fish stocks worldwide will collapse within 50 years.

How could this trend in fisheries be reversed?

Activity 17

Observe

Quick Code
ca8870s

Introducing Invasive Species

Did you know that many of the animals and plants you see every day are non-native? Some of the species are simply introduced/alien species, but others are known as invasive species. When species are invasive, they do not naturally occur in the area and they cause harm to the ecosystem. Invasive species can be introduced accidentally or deliberately. The images below show some common invasive species.

European Starling

Feral Cats

Jackson's Chameleon

Zebra Mussel

CCC **Stability and Change**

EXPLORE

Go online to explore some of California's invasive plant and animal species. Then, **answer** the questions below.

Which of the invasive species you learned about today have you seen before?

How could invasive species be a threat to biodiversity?

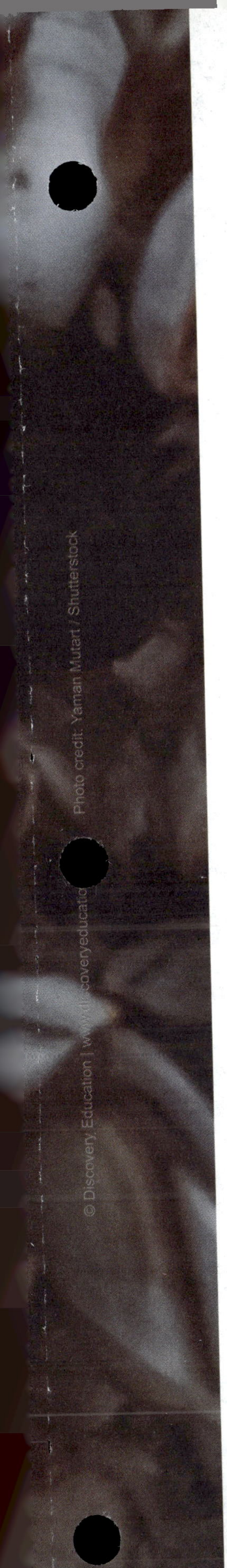

Activity 18

Analyze

Invasive Species in California

Quick Code
ca8871s

Work in pairs to **observe** the table below and **answer** the prompts that follow.

Species	Original Range	First Found in California	Damage to Biodiversity
European Green Crab	Europe and North Africa	Arrived most likely on a ship around 1990.	Eats many species of native shellfish and crabs.
Argentine Ant	Northern Argentina South America	Arrived around 1907. This ant has spread to all continents except Antarctica.	Attacks native ant species and disrupts native food webs. These ants also protect some insects that feed on crops.
Mute Swan	Western Europe	Deliberately released as an ornamental species.	These big birds eat submerged aquatic plants that provide shelter and food for native fish, wildfowl, and aquatic invertebrates.
Zebra Mussel	Europe	Arrived in 2008, probably via boats or irrigation canals bringing water from Colorado. Has spread across the United States.	Displaces native mussels from their habitat. Strains out plankton that form the base of freshwater food chains.

SEP **Analyzing and Interpreting Data**

List the different ways the invasive species in the table impact other organisms in their ecosystems.

What direct and indirect effects could invasive species have on biodiversity?

Watch this video, and then **answer** the question.

EXPLORE

How did the introduction of the zebra mussel affect food chains in the Hudson River?

Activity 19

Evaluate

Quick Code
ca8872s

Ocean Food Chains, Invasive Species, Effects on Food Webs, and No Asian Carp

Ocean Food Chains

What forms the base of the food chains as described in the video? **Circle** the correct answer.

- ☐ Algae that use sunlight to grow near the surface and nutrients brought up from the ocean floor by storms
- ☐ Nutrients from the wastes of ocean animals
- ☐ Plants that grow in the ocean and nutrients that wash in from the shore
- ☐ Plants, as in many land food chains

Invasive Species

Think about the interactions you saw in the video. How did the organisms interrelate? What do you predict might happen if someone introduced a new species that could eat large amounts of algae at a rapid pace, outcompeting other species already present? **Describe** what you think might happen, making sure to specify how at least three of the organisms in the video might be affected and give a detailed explanation for each organism. Make sure to **include** in your answer at least one organism that does not eat algae.

____

____

____

CCC Cause and Effect

CCC Stability and Change

EXPLORE

Effects on Food Webs

Think more about what might happen if an invasive exotic species that consumed large amounts of algae were introduced. Invasive species can have complex effects and may cause other species to increase or decrease. **Explain** one way in which a species of animal might increase in numbers if this happened.

No Asian Carp

Asian carp are invasive species in many parts of the United States, but have not yet been introduced into the waters of California. There are four different species, and they all grow to be quite large. They eat snails, mussels, aquatic vegetation, zooplankton, and algae, so they have the potential to disrupt the food chain just as zebra mussels do. However, unlike mussels, which are transported from place to place on boats and fishing gear, carp must swim up waterways to invade new areas. Many native fish migrate along the same waterways. **Describe** a solution to help keep Asian carp out of California. **Use** reasoning to **support** your ideas.

Activity 20

Analyze

Case Studies

Quick Code
ca8874s

Read about your assigned case study. **Write** the factors that led to the decline of the species and the ways in which the decline can be reversed in the following chart.

Causes of Decline	Possible Solutions

EXPLORE

SEP Asking Questions and Defining Problems

CCC Cause and Effect

Case Studies

Case Study: The Bald Eagle

It is hard to believe, but our national bird, the bald eagle, almost became extinct. Bald eagle numbers began to drop when the forests where eagles lived were cleared for farming and timber. Hunting reduced the number of prey species available to the bald eagle. At the same time, farmers shot bald eagles because they thought the eagles attacked their farm animals. Then, bald eagles began to die because of pesticide poisoning. Because bald eagles were at the top of their **food chain**, they accumulated many poisonous insecticides in their bodies. This caused them to lay eggs with very thin shells, which cracked before they hatched. This caused the bald eagle to become endangered. By 1963, there were only about 1,000 left in the United States.

The many threats to the bald eagle required many steps to ensure restoration. The killing of bald eagles was prohibited, their nest sites were protected, and the pesticides that once poisoned them were banned. Eagles were moved to areas where they had become locally extinct. Now, there are more than 10,000 bald eagles in the United States.

Case Study: Gray Wolf

You have already encountered the gray wolf as an example of a keystone species. At one time, around 2 million gray wolves could be found in the United States. As waves of immigrants came to and moved across the country, they converted natural habitats into farms. This brought them into conflict with gray wolves, primarily because the wolves sometimes attacked livestock. Humans waged war on the species, offering bounties to promote their eradication. By the early 1960s, only about 300 wolves remained in the deep woods of Minnesota.

The gray wolf was listed as an endangered species. Hunting, poisoning, and trapping of wolves was made illegal. Over the past 50 years, their population has increased to about 5,000 and some have been reintroduced to the Rockies in Yellowstone National Park. Wolves are top predators and need very large natural ecosystems in which to live. The national population of wolves is fragmented, with small populations often widely separated from other groups. These small populations lack genetic diversity. This lack of genetic diversity decreases the chances of these groups surviving. One group, on Isle Royale in the Great Lakes, has almost completely disappeared. Some livestock farmers want to start hunting wolves again. The future of the gray wolf in the lower 48 states is still not certain.

EXPLORE

If you are reading the bald eagle case study, **watch** the following video and use it to **add** information to your chart.

The Bald Eagle

If you are reading the gray wolf case study, **examine** the following image and use it to **add** information to your chart.

The Gray Wolf

Share your findings with your peers studying the other case study and **answer** the following questions.

What are some similarities and differences in the problems that caused each species to decline?

What are some similarities and differences in the solutions used to reverse that decline?

EXPLORE

Activity 21

Evaluate

Quick Code ca8875s

Elephant Tusks, Surviving to Reproduce, and Unnatural Natural Selection

Study the following graphs. Then, **answer** the questions that follow.

Variation in Tusk Length as a Function of Shoulder Height in African Elephants

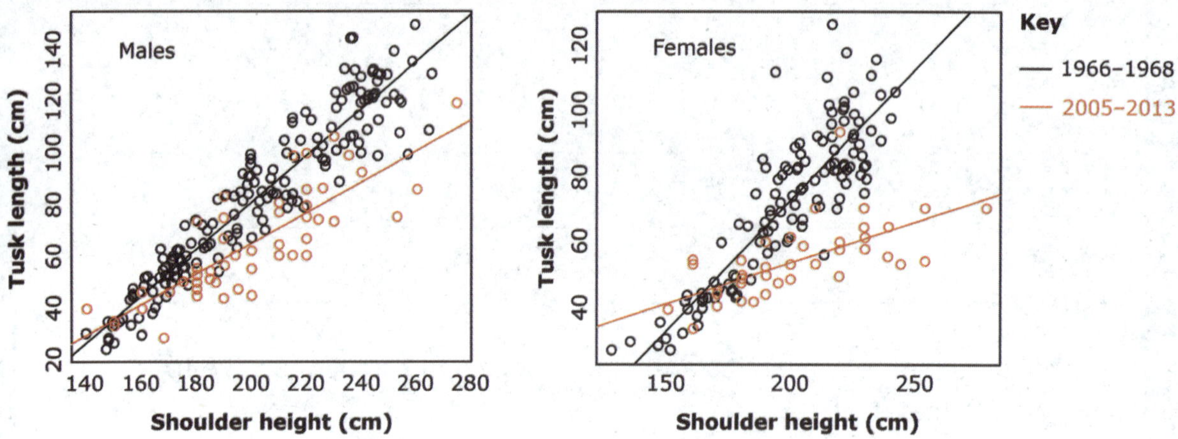

SEP Using Mathematics and Computational Thinking

CCC Patterns

CCC Cause and Effect

Elephant Tusks

What can you conclude about changes in elephant tusk lengths based on the graphs? **Justify** your conclusion with evidence from the data.

Surviving to Reproduce

Using the information in the graphs, **construct** a probability statement about how genetic variations of traits in a population increase some individuals' probability of surviving and reproducing in a specific environment. **Support** your statement with evidence.

Unnatural Natural Selection

Elephants use their tusks as offensive weapons in fending off predators. Using the information in the graphs, how do you predict the change in tusk length would impact elephant behavior over time? How do you predict the change will impact the rate of elephant harvesting to supply human ivory consumption? **Explain** your reasoning in making each prediction.

Activity 22

Record Evidence

The Eye of the Condor

Quick Code
ca8876s

As you worked through this lesson, you investigated and gathered evidence about biodiversity. Now, take another look at the image Eye of the Condor, which you first saw in Engage.

Let's Investigate the Eye of the Condor

How has your understanding of biodiversity and conservation changed?

SEP Engaging in Argument from Evidence

SEP Constructing Explanations and Designing Solutions

Read the Can You Explain? question from the beginning of this lesson.

Can You Explain?

What human activities harm biodiversity, and what human activities help to sustain biodiversity?

Use your new understanding of biodiversity and conservation to write a scientific explanation answering a question.

1. **Choose** a question. You may choose to answer the Can You Explain? question or one of the questions you wrote at the start of this lesson.
2. Use the following information to help you write your explanation.

EXPLAIN

What human activities harm biodiversity, and what human activities help to sustain biodiversity?

Talk Together

Share your scientific explanation with a partner. As you listen to their explanation, record the evidence they included and suggest opportunities for them to improve on their explanation.

Engaging in Argument
Record the writer's claim.
List three pieces of evidence the writer used to support the claim. Put a star next to evidence different from your scientific explanation.
Interpretation of the Facts – What facts were interpreted differently? Do you agree with the interpretation? Why or why not?

EXPLAIN

STEM in Action

Activity 23

Analyze

Quick Code
ca8906s

Becoming a Land-Use Planner

Read the passage and **highlight** details that help you understand what a land-use planner does.

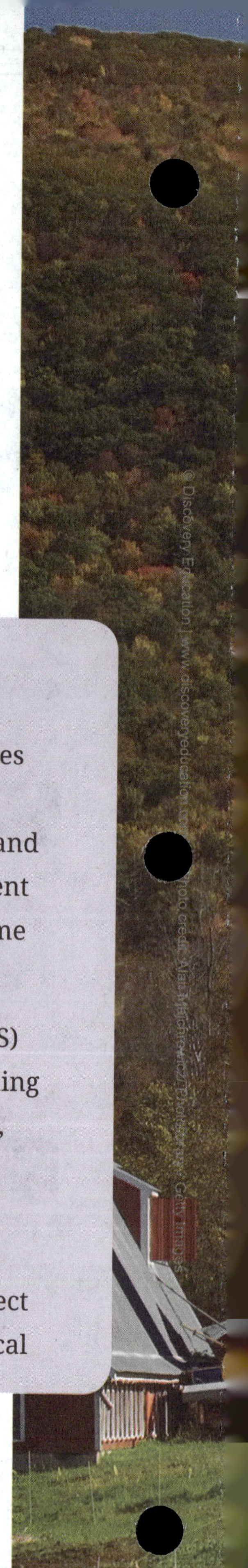

Becoming a Land-Use Planner

Human use of land can cause land degradation. Some human activities break down the land and cause problems for the future. They can destroy the soil and water. These practices can strip away both land and the organisms that live on and in it. Land-use planners work to prevent harmful effects of development. They especially try to prevent extreme changes, such as desertification of land.

An example of a land-use program is the U.S. Geological Survey (USGS) Navajo Land Use Planning Project. This project centers on land planning in the Navajo Nation. Dr. Margaret Hiza-Redsteer, a Native American, is one of the scientist leaders of this project. Together, the team has looked at how the land base of the reservation could grow to support the increasing human population. They needed to increase resources and develop the infrastructure to meet the increased needs. The project team developed a land-use plan, including information on geographical

SEP Asking Questions and Defining Problems

SEP Obtaining, Evaluating, and Communicating Information

hazards, bridges, and domestic septic and landfill systems. They also mapped out potentially available water sources. Specific sites were selected for detailed study to address the potential impact of planned changes. The team examined natural resource management, including farming projects, grazing practices, and vehicle use. The team also addressed any actions that could be linked to climate change and drought. Since the land in question had been subject to droughts, it was very important that any new land use would not make this situation worse.

Farm in Vermont

Land Degradation

ELABORATE

Read the passage and **highlight** details that show how planning land use protects the environment. Later, you will use what you learned to look at a proposal to build a new store adjoining an existing store on an empty lot in a suburban area. You will explain how as a land-use planner you can help humans and other species.

Land Use and Planning

The United States is home to over 300 million people today. A population of 300 million people can put quite a bit of stress on the land and environment. If you read or watch the news, you might see stories about crowded cities, suburban sprawl, or the loss of animal habitats. All of these issues are related. We cannot solve all of the problems that come with a large and growing population, but we can try to minimize our impact on the environment. Through careful planning and thoughtful use of the land, we can help prevent many of the problems affecting the world around us.

We must confront several issues related to how we currently use land. As a city grows, its boundaries are pushed outward, infringing on the land around the city. This sprawl can have serious consequences on the surrounding area. As vegetation is removed to make way for houses, the soil can be damaged, leading to more erosion and a greater likelihood of flooding. As the land use changes from rural to urban, native species can be forced out by construction. This can cause animals to die, or it can allow non native (invasive) species to thrive.

Sprawl can also eliminate entire ecosystems. When wetlands are lost, we lose many plants and animals, as well as the wetlands' ability to naturally filter water. One other problem with sprawl is the increased use

A large population can put stress on the environment.

of automobiles. When people live far from the center of a town or a city, they are more likely to drive to get the things they need. Increased driving leads directly to more pollution. This in itself causes many problems for the air, water, and human health.

Not to worry, though! The news is not all bad. Plenty of environmental problems are tied to the ways we use the land, but there are several ways we can reduce or eliminate these problems. All over the country, states, cities, and towns are working hard to keep land healthy. Let's explore a few of the methods people are using to prevent land-use problems.

One way to protect land is to require permits for all new development. For example, in many places, before people can build a house, a warehouse, or a store, they must file paperwork and pay a fee. City or county staff members then review this permit application and decide if the new construction is appropriate. If the development plan follows all of the guidelines, the project is allowed to continue. If the plan is deemed harmful, the project will be stopped. The fees that permit holders pay are meant to help offset some of the costs of construction. Permits help cities keep track of how the land is being used. Another method to protect land and prevent problems is zoning. Zoning is the process by which certain uses for land are allowed only in certain areas. Zoning also acts as a type of land-use restriction. For example, an area that has been zoned for residential use can only include houses. Large factories cannot be built in residential zones. The zoning regulations keep the pollution and noise of industry away from neighborhoods. Zoning can also require that some areas have buildings that are built close together. This can help concentrate, rather than spread out, the impact of that land use.

One other method to prevent environmental land problems is a conservation easement. A conservation easement helps protect open space or rural land by allowing a group of people (or a whole town) to buy a piece of rural land. The land is then conserved, and it can never be developed. Many times, farmers are still allowed to farm the land, but they may not build more houses on it. This keeps that rural land open and natural, even if a nearby city grows to reach the border of the land.

A conservation easement can help protect land from development.

Do you know how your area helps protect the land and prevent land-use problems? Can you think of any ways your town could better protect the land?

Land-Use Planner

Why would a land-use planner be important to a development project? In your answer, explain how the land-use planner helps both humans and other species.

ELABORATE

Planning a Store

Imagine that you are a land-use planner. You have been asked to help with a project to build a store on an empty lot next to another store in a suburban area. Describe at least three factors that could potentially impact the environment that need consideration as you begin work on this project. Choose one of these factors and propose a solution to limit the store's impact on the environment.

ELABORATE

Activity 24

Concept Review

Review: Sustaining Biodiversity

Quick Code ca8907s

Now that you have completed the objectives for this concept, **review** the core ideas you have learned. Record some of the core ideas below.

Core Ideas

EVALUATE

Talk with a Group

Now, think about the video Sensing Biodiversity from the Sky, which you saw in Get Started. Discuss how what you have learned about biodiversity can help you understand the role of technology in conservation efforts.

Solve Problems

Unit Project: Bringing Back from the Brink

Quick Code
ca8909s

Video

Park Ranger Holding Alligator Skin

SEP Obtaining, Evaluating, and Communicating Information

CCC Stability and Change

The video Elwha River Fish Restoration Plan provides an example of a human behavior that led to the extinction of a living process. This process involves the anadromous fish runs for salmon. The salmon runs have been extinct for about 100 years. However, recent actions have helped restore this process. Each problem threatening an organism has potential solutions, but they are often multifaceted, and restoration processes generally require changes in human behavior. In some cases, there are too many complications to provide a single solution that will solve the problem.

Elwha River Fish Restoration Plan

For this project, you will research information about an organism and the restoration programs that are currently in effect for that organism.

Reason

Species Restoration

In this investigation, you will **choose** an organism from a list provided or select a different organism with teacher approval. You will **conduct research** to learn about the organism's ecology, including information about its range, reasons for decline, current status, and any conservation efforts related to the organism. You will work in groups of three to **plan** a multimedia presentation or poster about your chosen organism for the capstone presentation.

What materials do you need? (per group)

- Paper
- Pens or pencils
- Tablets or computers
- Printer

Procedure

Select an organism and **plan** your presentation. **Record** the roles each group member will play, the type of presentation you will create, the information you want to include, and the areas of conservation you want to focus on.

Organism:	**My research role:**
Type of presentation:	
Information to include:	
Areas of conservation:	

Research

Research the organism and the conservation efforts. Record information about the organism's ecology (keystone species), range, reasons for decline, conservation plan, successes and failures, current status, and future prospects or problems. Then, create the presentation. Provide source information for all resources used to create the presentation.

Reflect

Why is it important to study an organism's ecology?

____

____

____

____

____

How do you think this information would be helpful in maintaining a species?

____

____

____

____

____

What information should you use when making suggestions for ways to improve existing programs?

Finding Sources

For your capstone project, you conducted research on an organism of your choice. Explain your research process. What search terms did you use, and how successful were these in helping you find information? How can you tell if your sources are credible?

Planning Document and Presentation

In the provided space, provide a plan for your presentation that includes your complete outline, sources, and type of presentation. Be prepared to share your presentation with your class.

Grade 8 **Glossary**

English — A — Español

English	Español
acceleration a change in the velocity (speed, direction, or both) of a body (related word: accelerate)	**aceleración** cambio en la velocidad (rapidez, dirección o ambas) de un cuerpo (palabra relacionada: acelerar)
air the part of the atmosphere closest to Earth; the part of the atmosphere that organisms on Earth use for respiration	**aire** parte de la atmósfera más cercana a la Tierra; la parte de la atmósfera que los organismos que habitan la Tierra utilizan para respirar
allele one member of a pair or series of genes on specific chromosomes in specific positions	**alelo** uno de los miembros de un par o serie de genes en cromosomas específicos en posiciones específicas
amber a fossil resin from tree sap, usually a translucent, yellow-brown color, which may contain insects or other organisms	**ámbar** resina fósil de un árbol, por lo general es translúcido, de color marrón amarillento; puede contener insectos u otros organismos.
amino acid one of the 20 types of molecules that combine to form proteins	**aminoácido** uno de los 20 tipos de moléculas que se combinan para formar proteínas
amplitude height or "strength" of a wave	**amplitud** altura o "fuerza" de una onda
antibiotic a chemical that kills or inhibits growth of bacteria	**antibiótico** un químico que mata o inhibe el crecimiento de bacterias
aquatic relating to water	**acuático** relativo al agua

Grade 8 **Glossary**

asteroid
a small object made of rock or metal that orbits the sun

asteroide
pequeño objeto de roca o metal que orbita alrededor del sol

---

atmosphere
the layers of gases that surround a planet

atmósfera
capas de gases que rodean un planeta

---

attract (magnetism)
the action between two opposing magnetic poles

atraer (magnetismo)
la acción entre dos polos magnéticos opuestos

---

axis
an imaginary line that an object spins or revolves around

eje
línea imaginaria sobre la cual gira o da vueltas un objeto

B

backbone
the column of bones that runs along the backs of all vertebrate animals and protects the spinal cord

espina dorsal
columna de huesos que se extiende a lo largo de la espalda de todos los animales vertebrados y protege la médula espinal

---

bacteria
single-celled organisms without an organized nucleus
(related terms: bacterium, prokaryote)

bacterias
organismos unicelulares sin un núcleo organizado (términos relacionados: bacteria, procariota)

---

biodiversity
the many different types of life that exist in an environment

biodiversidad
muchos y diferentes tipos de vida que existen en un medioambiente

---

biosphere
that part of Earth in which life can exist

biosfera
parte de la Tierra donde puede existir la vida

C

carbon
an element that is found in almost all compounds that make up living things

carbono
elemento que se encuentra en casi todos los compuestos que forman seres vivos

carbon dioxide
a waste product made by cells of the body; a gas in the air made of carbon and oxygen atoms: Humans rid themselves of carbon dioxide waste by exhaling, or breathing out.

dióxido de carbono
desecho formado por células del cuerpo; gas del aire formado por átomos de carbono y oxígeno: los humanos desechamos dióxido de carbono al exhalar o expulsar aire

cast
a type of fossil formed when sediment fills a mold of an organic object such as a fossil shell, creating a replica of that object made of sediment

calco
tipo de fósil formado cuando el sedimento llena un molde de un objeto orgánico, como una concha fósil y crea una réplica del objeto hecha de sedimento

Celsius
the metric temperature scale: Water boils at 100 degrees Celsius, and it freezes at 0 degrees Celsius

Celsius
escala de la temperatura métrica: el agua hierve a los 100 grados centígrados y se congela a los 0 grados centígrados

characteristic
a feature of an organism; something you can observe about an organism

característica
rasgo de un organismo; algo que se puede observar sobre un organismo

chlorophyll
a green pigment found in photosynthetic organisms

clorofila
pigmento verde que atrapa la energía del sol

climate
the average weather conditions in an area (related word: climatic)

clima
condiciones promedio del tiempo atmosférico en un área (palabra relacionada: climático)

Grade 8 **Glossary**

comet
a body of ice and dust that travels in a long narrow orbit around the sun

cometa
cuerpo de hielo y polvo que viaja en una órbita larga y estrecha alrededor del sol

---

conservation of energy
Energy cannot be created out of nothing, nor can it be completely destroyed. Energy can only change its form.

conservación de energía
La energía no se puede crear de la nada, y tampoco se la puede destruir por completo. La energía solo puede cambiar de forma.

---

contact
the act of touching. Two things are in contact when they are touching one another.

contacto
acto de tocar. Dos cosas están en contacto cuando se tocan entre sí.

---

core
the innermost layer of Earth, comprised of the liquid outer core and solid inner core; consists mainly of iron and nickel

núcleo
capa más interna de la Tierra, consta del núcleo externo líquido y el núcleo interno sólido; constituido principalmente por hierro y níquel

---

crater
a large, circular pit in the surface of a planet or other body in space usually formed when two bodies in space collide

cráter
hoyo grande y circular en la superficie de un planeta u otro cuerpo en el espacio, generalmente formado cuando chocan dos cuerpos en el espacio

D

decay
to break down or wear away

pudrirse
descomponerse o desgastarse

---

deciduous forest
a forest containing trees that lose their leaves each year

bosque caduco
bosque que contiene árboles que pierden las hojas cada año

---

deforestation
the clearing, by burning or logging, of trees in a forested area

deforestación
desmonte, por quema o tala, de árboles en un área forestal

density
an object's mass divided by its volume; a measure of how many particles are packed together into a certain amount of space (related word: dense)

densidad
masa de un objeto dividida por su volumen; medida que indica en qué forma se comprimen las partículas en cierta cantidad de espacio
(palabra relacionada: denso)

---

diameter
the length of a straight line that runs from one side of a circle (or sphere), through the middle of the circle (or sphere), to the other side; the width of a circle or sphere through its middle

diámetro
longitud de una línea recta que corre desde un lado de un círculo (o esfera), a través del medio del círculo (o esfera), hasta el otro lado; el ancho de un círculo o esfera a través de su medio

---

diffraction
the bending of light or sound waves around obstacles in its path

difracción
la desviación de ondas de luz o sonido alrededor de los obstáculos en su camino

---

disease
a condition that disrupts processes in the body

enfermedad
afección que perturba los procesos del cuerpo

DNA
found in the nucleus of a cell, a long nucleic acid molecule containing the genetic instructions for the development and functioning of all living organisms; abbreviation for deoxyribonucleic acid

ADN
molécula larga de ácido nucleico que se encuentra en el núcleo de una célula y que contiene las instrucciones genéticas para el desarrollo y funcionamiento de todos los organismos vivos; es la abreviatura de ácido desoxirribonucleico

---

drought
a prolonged shortage of rainfall

sequía
falta prolongada de lluvia

---

dwarf planet
a body in space that orbits the sun; similar to a small planet but without the proper mass

planeta enano
un cuerpo en el espacio que orbita alrededor del sol; similar a un planeta pequeño pero sin la masa adecuada

E

Earth the third planet from the sun; the planet on which we live (related words: earthly; earth - meaning soil or dirt)	**Tierra** tercer planeta desde el sol; planeta en el cual vivimos (palabras relacionadas: terrenal; tierra en el sentido de suelo o suciedad)
echo a sound wave that bounces off of a surface and reflects back to the source	**eco** onda sonora que rebota en una superficie y se refleja de regreso a la fuente
ecosystem all the living and nonliving things in an area that interact with each other	**ecosistema** todos los seres vivos y los elementos no vivos en un área que interaccionan entre sí
electric having to do with electricity; a form of energy commonly used for lights, electric motors, and heating	**eléctrico** que tiene relación con la electricidad; forma de energía comúnmente usada para luces, motores eléctricos y calor
electromagnet a metal object that acts as a magnet when an electric current moves through it	**electroimán** objeto de metal que actúa como un imán cuando una corriente eléctrica se mueve a través de él
electromagnetic spectrum the full range of frequencies of electromagnetic waves	**espectro electromagnético** rango completo de frecuencias de las ondas electromagnéticas
elliptical shaped like an oval	**elíptico** con forma ovalada

embryo
an early stage in the growth of an organism in which the organism is made up of a small number of cells and contained within a protective structure such as a seed or uterus

embrión
primera etapa en el desarrollo de un organismo en la cual el organismo está formado por un pequeño número de células y se encuentra en el interior de una estructura protectora, como una semilla o un útero

---

endangered species
a species that is in danger of extinction due to its low number of remaining individuals

especie en peligro de extinción
una especie que se encuentra en peligro de extinción debido a su bajo número de individuos restantes

---

energy
the ability to do work or cause change; the ability to move an object some distance

energía
habilidad para hacer un trabajo o producir un cambio; habilidad para mover un objeto a cierta distancia

---

environment
all the living and nonliving things that surround an organism

medio ambiente
todos los seres vivos y objetos sin vida que rodean a un organismo

---

equator
an imaginary line that divides Earth into Northern and Southern Hemispheres; located halfway between the North and South Poles
(related word: equatorial)

ecuador
línea imaginaria que divide la Tierra en Hemisferio Norte y Hemisferio Sur; ubicada a mitad de camino entre el Polo Norte y el Polo Sur
(palabra relacionada: ecuatorial)

---

evidence
data or information that either supports a statement or shows that it is not true

prueba
datos o información que respaldan un argumento o que demuestran que no es verdadero

---

evolution
A theory that the various types of animals and plants have their origin in other preexisting types and that the distinguishable differences are due to modifications in successive generations

evolución
teoría según la cual los diferentes tipos de animales y plantas provienen de otros tipos preexistentes; según esta teoría las diferencias distinguibles entre ellos se deben a los cambios que ocurren a través de generaciones sucesivas

extinct
describes a species of animals that once lived on Earth but which no longer exists (related word: extinction)

extinto
palabra que hace referencia a una especie de animales que una vez habitó la Tierra, pero que ya no existe
(palabra relacionada: extinción)

F

food chain
a model that shows one set of feeding relationships among living things

cadena alimentaria
modelo que muestra un conjunto de relaciones alimentarias entre seres vivos

food web
a model that shows many different feeding relationships among living things

red alimentaria
modelo que muestra muchas y diferentes relaciones de alimentación entre los seres vivos

force
a pull or push that is applied to an object

fuerza
acción de atraer o empujar que se aplica a un objeto

fossil
evidence that an organism once existed in an area; can be a part of the organism's body or a trace fossil which is a mark or print left by the organism (related word: fossilize)

fósil
muestra de que un organismo existió una vez en un área; puede ser una parte del cuerpo del organismo o un rastro fósil, que es una marca o impresión dejada por el organismo
(palabra relacionada: fosilizar)

fossil fuels
various types of nonrenewable fuels naturally formed from the remains of plants and animals that died thousands or millions of years ago

combustibles fósiles
varios tipos de combustibles formados de manera natural a partir de los restos de plantas y animales que murieron hace miles o millones de años

fossil record
the information about Earth's history that can be gathered from fossils

registro fósil
información sobre la historia de la Tierra que puede reunirse a partir de los fósiles

frame of reference
an item against which the motion of an object can be measured

sistema de referencia
elemento o elementos respecto a los cuales puede medirse el movimiento de un objeto

---

frequency
the number of cycles a wave completes in a period of time; the number of times something happens in a period of time

frecuencia
cantidad de ciclos que completa una onda en un período; cantidad de veces que sucede algo en un período

---

friction
a force that opposes the motion of a body across a surface or through a gas or liquid

fricción
fuerza que se opone al movimiento de un cuerpo sobre una superficie o a través de un gas o un líquido

---

fusion
a nuclear reaction in which nuclei combine to produce more massive nuclei resulting in the release of large amounts of energy

fusión
reacción nuclear donde los núcleos se combinan para producir más núcleos masivos, que resultan en la liberación de grandes cantidades de energía

G

galaxy
a group of solar systems, dust, and gas held together by gravity; our solar system is part of the Milky Way galaxy

galaxia
grupo de sistemas solares, polvo y gas unidos por la gravedad; nuestro sistema solar es parte de la galaxia llamada Vía Láctea

---

gas
a state of matter without any defined volume or shape in which atoms or molecules move about freely

gas
estado de la materia sin volumen ni forma definidos en el cual los átomos o moléculas se mueven libremente

---

gene
the basic unit of heredity in a living organism; a segment of DNA or RNA

gen
unidad básica de la herencia en un organismo vivo; segmento de ADN o ARN

---

genetic variation
range of differences in DNA among organisms

variación genética
rango de diferencias en el ADN entre organismos

Grade 8 Glossary

germs
disease-causing microorganisms

gérmenes
microorganismos que causan enfermedad

glacier
a large mass of ice resting on, or overlapping, a land surface

glaciar
una gran masa de hielo que descansa sobre o se superpone con una superficie terrestre

gravity
a force that exists between any two objects that have mass (related word: gravitational)

gravedad
fuerza que existe entre dos objetos cualquiera que tienen masa
(palabra relacionada: gravitacional)

greenhouse gas
a gas, usually carbon-based, that contributes to global warming through the greenhouse effect, which prevents the escape of radiant heat from Earth's atmosphere

gas invernadero
gas, por lo general a base de carbono, que contribuye al calentamiento global mediante el efecto invernadero, el cual impide que el calor radiante salga de la atmósfera terrestre

groundwater
water stored below Earth's surface in soil and rock layers

agua subterránea
agua almacenada por debajo de la superficie de la tierra, en capas de suelo y rocas

H

heat
the transfer of thermal energy

calor
transferencia de energía térmica

hertz
measurement of wave frequency

hercio
medida de la frecuencia de onda

hydrogen
the chemical element consisting of one proton and one electron

hidrógeno
elemento químico que consiste en uno o más protones y un electrón

I

infrared
electromagnetic energy that has a lower frequency than visible light and is beyond the color of red on the light spectrum: Infrared light is invisible to humans

infrarrojo
energía electromagnética que tiene una frecuencia menor que la luz visible y que está más allá del color rojo en el espectro de luz: la luz infrarroja es invisible para los seres humanos

inherit
to receive genetic information and traits from a parent or parents
(related word: inheritance)

heredar
recibir información y rasgos genéticos de un padre o de los padres
(palabra relacionada: herencia)

interference
the effect of two or more waves traveling through the same medium

interferencia
efecto de dos o más ondas que viajan por el mismo medio

K

keystone species
within the ecological community, this species has a critical role in maintaining the structure of the community

especies clave
dentro de la comunidad ecológica, estas especies tienen un papel crucial en el mantenimiento de la estructura de la comunidad

kinetic energy
the energy an object has due to its motion

energía cinética
energía que tiene un objeto debido a su movimiento

kinetic friction
a resistive force between two objects moving pass each other

fricción cinética
fuerza de resistencia entre dos objetos que se mueven uno sobre el otro

Kuiper Belt
an area in the solar system beyond the planets where many small, icy bodies are located

Cinturón de Kuiper
área del sistema solar que se encuentra más allá de los planetas, donde se ubican muchos cuerpos pequeños y helados

L

latitude
angular distance north and south of the equator

latitud
distancia angular al norte y sur del ecuador

light
waves of electromagnetic energy; electromagnetic energy that people can see

luz
ondas de energía electromagnéticas; energía electromagnética que la gente puede ver

light year
the distance light travels in a vacuum in one year; about 6 trillion miles

año luz
distancia que viaja la luz en el espacio en un año; alrededor de 6 billones de millas

liquid
a state of matter with a defined volume but no defined shape and whose molecules roll past each other

líquido
estado de la materia con un volumen definido pero sin forma definida y cuyas moléculas se deslizan unas sobre otras

longitudinal waves
a wave that occurs when the particles of a medium move parallel to the direction of the wave

onda longitudinal
onda que se produce cuando las partículas de un medio se mueven en paralelo a la dirección de la onda

M

magnet
an object with a north and south pole that produces a magnetic field
(related term: magnetism, magnetic)

imán
objeto con un polo norte y un polo sur que produce un campo magnético
(palabra relacionada: magnetismo, magnético)

magnetic field
a region in space near a magnet or electric current in which magnetic forces can be detected

campo magnético
región en el espacio cerca de un imán o de una corriente eléctrica, donde pueden detectarse fuerzas magnéticas

magnify
to make something appear larger, usually by using one or more lenses

ampliar
hacer que algo parezca más grande, generalmente usando una o más lentes

---

magnitude
size of ground movement caused by seismic waves due to energy released

magnitud
tamaño del movimiento de la tierra provocado por las ondas sísmicas debido a la energía liberada

---

mangrove
tree that grows in tidal zones

manglar
árbol que crece en zonas costeras

---

mass
the amount of matter in an object

masa
cantidad de materia en un objeto

---

matter
material that has mass and takes up some amount of space

materia
material que tiene masa y ocupa espacio

---

measure
to use a tool to learn more about the volume, length or weight of an object
(related word: measurement)

medir
usar una herramienta para saber más sobre el volumen, la longitud o el peso de un objeto
(palabra relacionada: medición)

---

meteor
a streak of light in the sky that forms when a piece of rock from space burns up in Earth's atmosphere (related term: shooting star)

meteoro
reflejo de luz en el cielo que se forma cuando una pieza de roca del espacio se quema en la atmósfera de la Tierra
(palabra relacionada: estrella fugaz)

---

mold
a fossil that forms when the remains of an organism leave an imprint in the sediment after the organic material has been completely replaced

molde
fósil que se forma cuando los restos de un organismo dejan una impresión en el sedimento después de que el material orgánico haya sido remplazado completamente

moon a body in outer space that orbits a planet; a natural satellite	**luna** cuerpo en el espacio exterior que orbita alrededor de un planeta; satélite natural
motion a change in the position of an object compared to another object (related term: move, movement)	**movimiento** cambio en la posición de un objeto en comparación con otro objeto (palabra relacionada: mover, desplazamiento)

N

nebulae an interstellar cloud made up of hydrogen gas, plasma, helium gas, and dust	**nebulosa** una nube interestelar constituida por gases (hidrógeno y helio), plasma y polvo
nebular hypothesis the theory that a solar system evolves from a hot gaseous nebula	**hipótesis nebular** teoría que sostiene que un sistema solar evoluciona a partir de una nebulosa gaseosa caliente
Newton's laws the three fundamental laws Isaac Newton discovered governing the motion of objects	**leyes de Newton** las tres leyes fundamentales que Isaac Newton descubrió sobre el movimiento de los objetos
nitrogen an element that makes up most of the air near Earth's surface: nitrogen is a gas at room temperature	**nitrógeno** elemento que forma la mayor parte del aire que se encuentra cerca de la superficie de la Tierra: el nitrógeno es un gas a temperatura ambiente

O

offspring a new organism that is the product of reproduction	**descendencia** organismo nuevo que es el producto de la reproducción

optical having to do with the eye or lenses	**óptico** relacionado con los ojos o con las lentes
orbit the circular path of an object as it revolves around another object	**órbita** trayectoria circular de un objeto que se forma a medida que gira alrededor de otro objeto.
organism any individual living thing	**organismo** todo ser vivo individual
outer core the liquid outer portion of Earth's core, composed primarily of iron and nickel	**núcleo externo** parte exterior, líquida, del núcleo de la Tierra, compuesto principalmente por hierro y níquel
oxygen an element that makes up about 21% of Earth's atmosphere; a gas in Earth's atmosphere and in water that living organisms breathe	**oxígeno** elemento que forma aproximadamente el 21% de la atmósfera de la Tierra; gas que se encuentra en la atmósfera de la Tierra y en el agua que los organismos vivos respiran

P

photosynthesis the process in which plants and some other organisms use the energy in sunlight to make food	**fotosíntesis** proceso en el cual las plantas y algunos otros organismos usan la energía del Sol para producir alimentos
planet a large circular mass that revolves around a star	**planeta** gran masa circular que gira alrededor de una estrella
plate tectonics theory the theory that Earth's crust is made up of individual plates that gradually move in relation to each other	**teoría de la tectónica de placas** teoría que dice que la corteza de la Tierra está formada por placas individuales que se mueven gradualmente en relación una con otra

Grade 8 **Glossary**

polar
in earth science, having to do with the areas on Earth closest to the geographic North or South poles; in chemistry, describes a molecule that has a positively charged side and a negatively charged side
(related words: pole, polarity)

polar
en la ciencia de la tierra, tiene que ver con las áreas de la Tierra más próximas a los polos geográficos Norte y Sur; en química, describe una molécula que tiene un lado con carga positiva y otro con carga negativa
(palabra relacionada: polo, polaridad)

---

pole
the opposite ends of a battery, a magnet, or the north and south ends of Earth

polo
extremos opuestos de una batería eléctrica, un imán, o los extremos norte y sur de la Tierra

---

pollution
harmful materials put into the air, water, or soil at a faster rate than the environment can be cleansed
(related words: pollute, pollutant)

contaminación
materiales perjudiciales que se depositan en el aire, el agua o el suelo a un ritmo más rápido del que se pueden eliminar del ambiente
(palabras relacionadas: contaminar, contaminante)

---

population
the group of organisms of the same species living in the same area

población
grupo de organismos de la misma especie que viven en la misma área

---

position
a specific point in space

posición
punto específico en el espacio

---

potential energy
the amount of energy that is stored in an object; energy that an object has because of its position relative to other objects

energía potencial
cantidad de energía almacenada en un objeto; energía que tiene un objeto debido a su posición relativa con otros objetos

---

precipitation
water that is released from clouds in the sky; includes rain, snow, sleet, hail, and freezing rain

precipitación
agua liberada de las nubes en el cielo; incluye la lluvia, la nieve, la aguanieve, el granizo y la lluvia congelada

---

protein
a nutrient used by animals

proteína
nutriente usado por los animales

R

reflect
to strike a surface and bounce back in the opposite direction (related word: reflection)

reflejar
golpear sobre una superficie y rebotar en la dirección opuesta
(palabra relacionada: reflexión)

---

refraction
a change in the direction and velocity which occurs when a seismic or electromagnetic wave travels from one material into another of different density, state, or elasticity

refracción
cambio en la dirección y velocidad que se produce cuando una onda sísmica o una onda electromagnética pasan de un material a otro que tiene diferente densidad, estado o elasticidad

---

repel
to force an object away or to keep it away
(related term: repulsion)

repeler
forzar a un objeto para que se aleje o mantenerlo alejado (término relacionado: repulsión)

---

revolution
the orbiting of an object around another object

revolución
movimiento por el cual un objeto gira alrededor de otro objeto describiendo una órbita completa

---

rotate
turning around on an axis; spinning
(related word: rotation)

rotar
girar sobre un eje; dar vueltas
(palabra relacionada: rotación)

---

rotation
the spinning of a celestial body, such as a planet, around an axis

rotación
giro de un cuerpo celeste, como un planeta, alrededor de un eje

S

satellite
a natural or artificial object that revolves around another object in space

satélite
objeto natural o artificial que gira alrededor de otro objeto en el espacio

sedimentary rock a type of rock formed by gravity pressing fragments of other rocks and minerals together as they settle on land or under the ocean over a long period of time	**roca sedimentaria** tipo de roca que se forma cuando la gravedad junta los fragmentos de otras rocas y minerales a medida que se asientan en la tierra o debajo del océano, a través de un largo período
solar energy energy that comes from the sun	**energía solar** energia que proviene del Sol
solar system a system of objects that revolve around a star	**sistema solar** conjunto de objetos que giran alrededor de una estrella
solid matter with a fixed volume and shape	**sólido** materia con un volumen y una forma determinada
sound a vibration that travels through a material, such as air or water; something that you sense with your hearing	**sonido** vibración que viaja a través de un material, como el aire o el agua; lo que se percibe a través de la audición
sound wave a sound vibration as it is passing through a material: Most sound waves spread out in every direction from their source.	**onda sonora** vibración de sonido que se produce cuando se atraviesa un material: La mayoría se dispersa desde la fuente en todas direcciones.
space the large, empty part of the universe that does not contain any matter	**espacio** parte vacía y enorme del universo que no contiene ningún tipo de materia
species a group of organisms that share similar characteristics and can mate with each other to produce offspring	**especie** grupo de organismos que comparten características similares y que se pueden aparear para generar descendencia

speed distance traveled per unit of time	**rapidez** distancia recorrida por unidad de tiempo
star a massive ball of gas in outer space that gives off heat, light and other forms of radiation	**estrella** bola masiva de gas en el espacio exterior que emite calor, luz y otras formas de radiación
sun any star around which planets revolve	**sol** toda estrella alrededor de la cual giran los planetas
surface the top of an object; the outside of an object; the boundary between two objects or materials	**superficie** parte superior de un objeto; exterior de un objeto; límite entre dos objetos o materiales
sustainable describes a material or resource that is able to meet the demands of current use and yet be maintained in usable quantities to meet indefinite future demands	**sostenible** describe un material o recurso que es capaz de satisfacer las demandas de uso actual y mantenerse en cantidades utilizables para satisfacer demandas futuras indefinidas
system a group of parts that work together to function or perform a task	**sistema** un grupo de partes que trabajan juntas para funcionar o realizar una tarea

T

telescope an instrument used to observe objects that are far away	**telescopio** instrumento usado para observar objetos que se encuentran alejados
The Theory of Evolution A theory that the various species of living organisms have their origin in common ancestors and that the distinguishable differences are due to heritable modifications in successive generations	**teoría de la evolución** teoría de que varias especies de organismos vivos tienen su origen en antepasados comunes y que las diferencias que se distinguen se deben a modificaciones heredadas en generaciones sucesivas

theory
a set of statements or principles devised to explain a group of facts or phenomena, especially one that has been repeatedly tested or is widely accepted and can be used to make predictions about natural phenomena

teoría
conjunto de declaraciones o principios concebido para explicar un grupo de hechos o fenómenos, especialmente uno repetidamente probado o ampliamente aceptado y que puede usarse para hacer predicciones sobre fenómenos naturales.

---

thrust
a reaction force that pushes an object forward

empuje
fuerza de reacción que impulsa a un objeto hacia adelante

---

topography
detailed mapping of the physical features of a small locale or area

topografía
características físicas que definen el relieve de un lugar, como montañas, valles y la forma de los accidentes geográficos

---

trait
a characteristic or property of an organism

rasgo
característica o propiedad de un organismo

---

translucent
describes materials that allow some light to travel through them, but not enough to see through the material

translúcido
describe materiales que permiten que pase algo de luz a través de ellos, pero no lo suficiente como para ver a través del material

---

transparent
describes materials through which light can travel; materials that can be seen through

transparente
describe materiales a través de los cuales puede viajar la luz; materiales a través de los cuales se puede ver

---

transverse waves
a wave that occurs when the particles of a medium are displaced perpendicularly to the direction of the wave

onda transversal
onda que se produce cuando las partículas de un medio se mueven perpendicularmente a la dirección de la onda

tundra
extremely cold climate located near the North and South Poles and on the tops of mountains; receives very little precipitation and has no trees

tundra
clima extremadamente frío ubicado cerca de los polos Norte y Sur y en la cima de las montañas; recibe muy pocas precipitaciones y no tiene árboles

U

universe
everything that exists in, on and around Earth

universo
todo lo que existe en, sobre o alrededor de la Tierra

V

vacuum
an area in which there is no matter

vacío
área donde no hay materia

velocity
the speed and direction of moving objects

velocidad
la rapidez y dirección de objetos en movimiento

vertebrae
the individual bones that stack to form the backbone of an animal
(related word: vertebra)

vértebras
huesos individuales que se apilan para formar la columna vertebral de un animal

vibration
the rapid movement of an object back and forth

vibración
rápido movimiento de un objeto hacia delante y hacia atrás

virus
a sub-microscopic infectious agent that is able to grow or reproduce only within a host cell

virus
agente infeccioso submicroscópico que puede crecer o reproducirse solamente dentro de una célula huésped

W

water
a compound made of hydrogen and oxygen

agua
compuesto formado por hidrógeno y oxígeno

water cycle
the continual movement of water between the land, ocean, and the air

ciclo del agua
movimiento continuo del agua entre la tierra, los océanos y el aire

water vapor
the gaseous form of water; produced when water evaporates

vapor de agua
estado gaseoso del agua; se produce cuando el agua se evapora

wave
a disturbance caused by a vibration; Waves travel away from the source that makes them.

onda
perturbación causada por una vibración que se aleja de la fuente que la forma

wavelength
the distance between one peak and the next on a wave

longitud de onda
distancia entre un pico y otro en una onda

weather
the properties of the atmosphere at a given time and location, including temperature, air movement and precipitation

tiempo atmosférico
propiedades de la atmósfera en un determinado momento y lugar; entre ellas, la temperatura, el movimiento de aire y las precipitaciones

weight
the force of gravity on an object

peso
fuerza de gravedad que se ejerce sobre un objeto

wetland
an area of land where there is water on the surface or where the soil is completely filled with water for at least part of the year

humedal
área de tierra donde hay agua en la superficie o donde el suelo está completamente lleno de agua durante, por lo menos, parte del año

Index

A

B

C

D

E

N

O

P

R

S

T